IET MATERIALS, CIRCUITS AND DEVICES SERIES 35

Design of Terahertz CMOS Integrated Circuits for High-Speed Wireless Communication

Other volumes in this series:

Design of Terahertz CMOS Integrated Circuits for High-Speed Wireless Communication

Minoru Fujishima and Shuhei Amakawa

The Institution of Engineering and Technology

Published by The Institution of Engineering and Technology, London, United Kingdom

The Institution of Engineering and Technology is registered as a Charity in England & Wales (no. 211014) and Scotland (no. SC038698).

The Institution of Engineering and Technology
Michael Faraday House
Six Hills Way, Stevenage
Herts, SG1 2AY, United Kingdom

www.theiet.org

British Library Cataloguing in Publication Data
A catalogue record for this product is available from the British Library

ISBN 978-1-78561-387-6 (hardback)
ISBN 978-1-78561-388-3 (PDF)

Typeset in India by MPS Limited
Printed in the UK by CPI Group (UK) Ltd, Croydon

Contents

Preface

This book covers most of our current knowledge, from the background to the design of a terahertz complementary metal oxide semiconductor (CMOS) transceiver for realizing ultrahigh-speed wireless communication. The content of the book is mainly the result of the 5-year project of the Japanese Ministry of Internal Affairs on the 300-GHz-band transceiver with the silicon CMOS-integrated circuits from 2014 to 2019, with its related technology. Expected readers are not only CMOS-integrated circuit designers but also all those who are interested in ultrahigh frequency circuits. Many hints for developing ultrahigh frequency circuits on integrated circuits are described in this book. Although the content of this book is technical, readers are able to understand the contents with general knowledge of microwave textbooks. The target of the reader is graduate-student level specialized in electronics.

Main area of this book is wireless communication. Radio waves need to be coherent to realize efficient wireless communication. Radio waves are not necessarily coherent, on the other hand, in a sensor comprising a light source and a light receiver. As a result, the technical difficulty of the coherent communication circuit is higher than that of the sensor circuit. Since it is academically attractive to solve this difficult problem, the communication circuit was chosen as a research target. Note, however, that circuit techniques discussed in this book can also be applied to sensor circuits.

On the other hand, it is necessary to discuss not only the academic significance but also engineering significance of terahertz communication. Chapter 1 outlines the reasons for using terahertz for wireless communication and the current research on terahertz transceiver. The largest challenge of realizing terahertz communication with a CMOS-integrated circuit is how to realize a communication circuit exceeding the maximum oscillation frequency, $f_{\max}$, of the transistor. Chapter 2 discusses the theory and design of ultrahigh frequency amplifiers, and the layout for CMOS-integrated circuits, which are most important for realizing ultrahigh-speed circuits. Prior to this project, the knowledge obtained by other national projects on millimeter-wave integrated circuits we have been working on has greatly helped design terahertz CMOS circuits. Chapter 3 describes individual technologies that are important in designing ultrahigh-frequency CMOS-integrated circuits including terahertz. A terahertz transceiver is realized while utilizing these techniques. Chapter 4 describes several types of circuits and modules of terahertz CMOS transceivers. The evaluation method is also discussed. Finally, Chapter 5 discusses the future of terahertz communication.

The authors would like to thank a number of people who supported this project. In particular, Prof. Mizuki Motoyoshi (currently Tohoku University), Prof. Kyoya Takano (currently Tokyo University of Science), Prof. Kosuke Katayama (currently

Waseda University), Prof. Sang-yeop Lee, Dr. Ruibing Dong of Hiroshima University, and Dr. Shinsuke Hara of National Institute of Information and Communications Technology (NICT) are core members of research on terahertz circuit design and evaluation. In module design, Dr. Koichi Mizuno (currently Nagoya University) and Mr. Junji Sato of Panasonic greatly helped us. In addition, in the research planning, we made fruitful discussions with Dr. Kazuaki Takahashi of Panasonic, Dr. Iwao Hosako and Dr. Akifumi Kasamatsu of NICT, and Prof. Takeshi Yoshida of Hiroshima University. Besides, no research results described in this book could be obtained without support from many people including secretaries and students. We would like to take this opportunity to express our appreciation to all.

Chapter 1
Introduction

1.1 Terahertz communication

Why do we need to study terahertz communication? Will something great happen if "terahertz" and "communication" are blended? Let us start discussions on these two questions that an ordinary person probably wonders. This book covers a complementary metal oxide semiconductor (CMOS) transceiver with a frequency range of about 300 GHz, which is promising for terahertz communication, and the most important technical area for designing a circuit in the frequency range. Although this frequency band is terahertz, it is also partially included in millimeter waves. Therefore, first, millimeter waves and terahertz are briefly introduced, and it will be discussed why this frequency is promising for communication. Then possible applications will be discussed from the nature of terahertz communication.

Figure 1.1 shows the difference between commonly used microwave and millimeter wave used in still limited application. For example, the wireless local area network (LAN) uses 2.4 or 5 GHz. The wavelengths in the atmosphere are 12.5 and 6 cm, respectively. On the other hand, millimeter waves are radio waves whose wavelengths in the atmosphere are 1–10 mm as the name implies. Therefore, the wavelength is 1/10 to 1/100 as compared with general microwave. Frequencies are 10–100 times

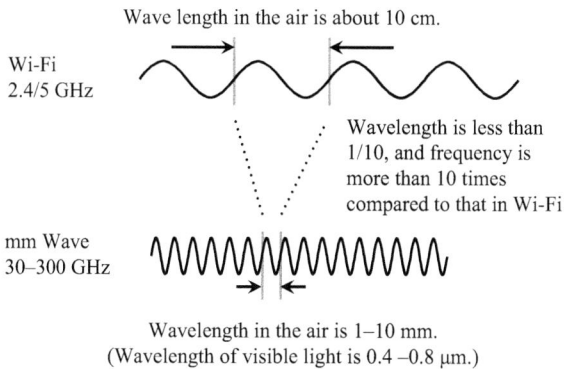

Wave length in the air is about 10 cm.

Wi-Fi
2.4/5 GHz

Wavelength is less than
1/10, and frequency is
more than 10 times
compared to that in Wi-Fi

mm Wave
30–300 GHz

Wavelength in the air is 1–10 mm.
(Wavelength of visible light is 0.4 –0.8 μm.)

Figure 1.1   What is millimeter wave? Considering from the difference from microwave

higher than microwave. Since the wavelength of visible light is 0.4–0.8 μm, millimeter waves are apparently close to microwave. However, its nature is approaching the light. This is important in considering the application, which will be discussed later.

The frequency band of the millimeter wave is clearly defined, and it is from 30 to 300 GHz. Strictly speaking, since the speed of light in vacuum is only slightly slower than 300,000 km/s, the frequency could shift slightly higher from 30 to 300 GHz if we define the wavelength as 1–10 mm. However, millimeter waves are not defined by wavelength but by frequency as shown in Figure 1.2. On the other hand, the definition of terahertz is ambiguous and is a frequency around 1 THz. In the narrow definition, the frequency range of 0.3–3 THz is recognized as terahertz, which is also called submillimeter wave with wavelengths from 0.1 to 1 mm. On the other hand, 0.1–10 THz can be regarded as terahertz if it is near 1 THz. The frequency of 300 GHz, which is the main target in this book, belongs indeed to this ambiguous area. Therefore, the terahertz technique discussed in this book may apply to millimeter waves as well. Although 0.1–0.3 THz is sometimes called millimeter wave, in this book, frequency near 300 GHz is called as terahertz.

Well, why do we need to study terahertz? The largest motivation is, indeed, to exploit resources. However, even if it is claimed that the aim is to exploit the resources, we will not be able to immediately understand what we are saying. Let us first discuss what resources are. Resources are "things" useful for various activities in human life and industry. Resources include natural resources and nonnatural resources. As shown in Figure 1.3, various resources exist. Radio waves are recognized as one of the resources. Radio waves are categorized in nonnatural resources, because people will produce electric waves except for the radio waves that pour down from space. But why do we need to consider radio waves as resources?

To understand radio waves as resources, it is useful to know the nature of radio waves with the schematic diagram shown in Figure 1.4. Basically, radio waves of the same frequency in the same place at the identical time can be used by only one person. When multiple people use the same frequency at the same place at the identical time,

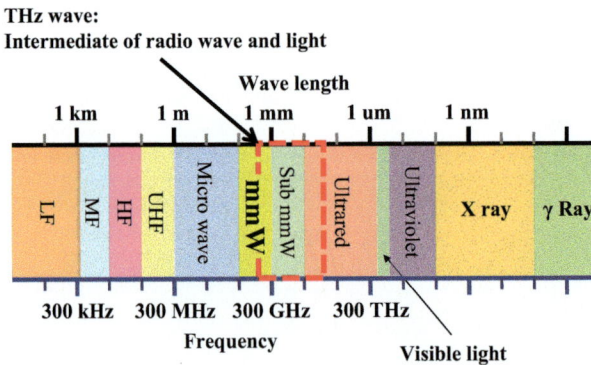

Figure 1.2   What are millimeter wave and terahertz? Let us see the relationship between millimeter waves and terahertz in electromagnetic waves

Example

Natural resource Used in materials existing in nature	Water resource Mineral resource Forest resource Aquatic resource Undersea resource

Resource
Used for various
activities such as
human life and
industry

Nonnatural resource Used with substances added with some artificial action	Tourism resources Radio wave resources Human resources

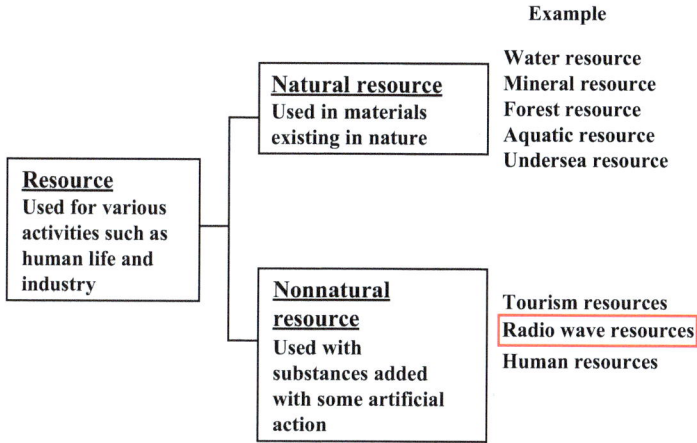

Figure 1.3　What are resources? The first purpose of studying terahertz communication is to exploit radio resources

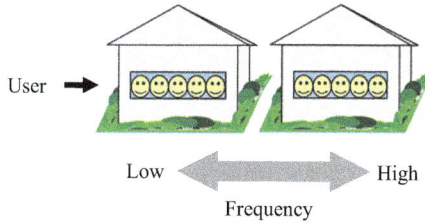

User →

Low ⟷ High

Frequency

Figure 1.4　The relationship between the radio resources and the frequency is like the relationship between the living place and the person

the radio waves mix in the atmosphere and crosstalk occurs. Strictly speaking, when radio waves are scrambled with specific codes as in CDMA or individual communications are realized with a narrowed radio beam, crosstalk may not occur at the same frequency at the same place in the same time. Crosstalk will occur, however, unless exceptional cares are paid. Since electromagnetic (EM) waves continuously exist from radio waves to light, available frequency resources exist infinitely in theory. In practice, however, the upper limit of the available frequency is determined by the technology limitation. In a situation where a specific user occupies a certain frequency while the frequency range is limited, it is necessary to consider which frequency should be assigned to whom. That is why radio waves need to be regarded as finite resources.

Figure 1.5 shows the current issue on radio wave resources and the research directions to solve it. In the current state, because existing users already occupy available frequency resources, no one can start new services even if he has ideas and passion.

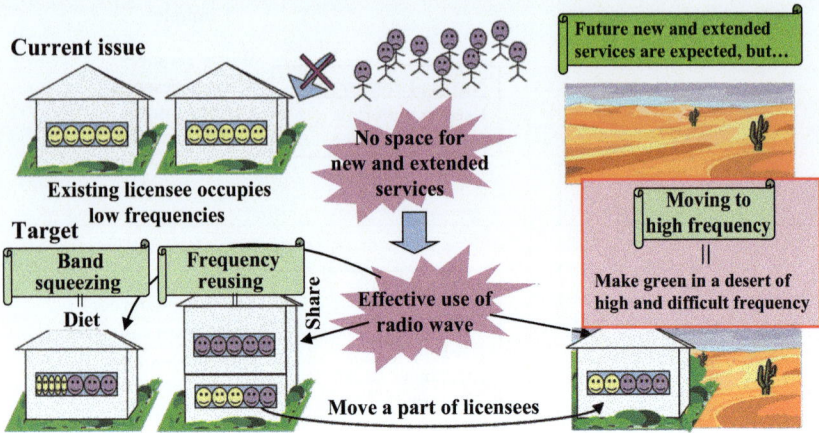

Figure 1.5   Schematic diagram of research to explore radio resources. There are several studies on the development of radio wave resources, but exploring high frequencies is essentially important

Accordingly, development of radio wave resources is indispensable for future-service development. There are several ways to develop radio resources. The first is to reduce the frequency bandwidth required for transmitting the same information. For example, even when sending the same amount of information, it is possible to reduce the required frequency bandwidth by changing the modulation from analog to digital, using symbols containing many digital information, and/or compressing digital information. In 2009, analog TV broadcasting in Japan switched to digital broadcasting. As a result, the total frequency bandwidths occupied by TV broadcasting are reduced, and the 700-MHz band has been released. The second is to reuse frequencies. Here, a method called cognitive radio is used. Even if a certain frequency is assigned to a specific user, the service is not always provided using that frequency. If the time, during which the allocated frequency is unused, can be cognized, it is possible for another user to use that frequency in only that period. This method also increases the number of users. The last is to increase the upper limit of available frequencies. However, the higher the frequency, the dramatically higher the technical difficulty to use that frequency. Although this is the major factor preventing the use of high frequencies, most issues can be solved by technologies unless theoretically impossible. If the upper limit of the available frequencies rises, radio resources will increase. This is the largest incentive to study terahertz.

However, from a user's point of view, people will not be interested in which frequency to use. Instead, people are interested in the benefits to be provided. The most obvious benefit of using high frequencies is improving the data rate. Generally, as shown in Figure 1.6, communication data rates using low frequency and high frequency are different. Of course, using higher frequencies can send higher data rates. In other words, the data rate overwhelmingly increases as the frequency rises.

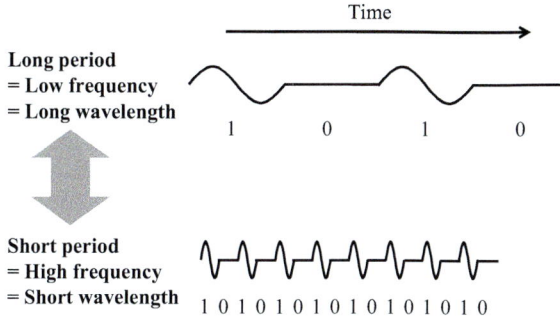

Figure 1.6 Radio wave with short period (high frequency) can transmit more information in the same time

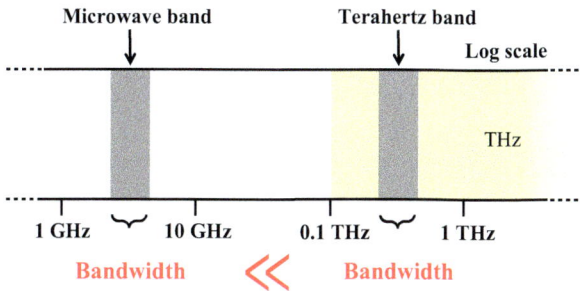

Figure 1.7 Communication capacity in terahertz is much larger than in microwave according to Shannon–Hartley theorem

Let us theoretically discuss the relationship between frequency and data rate. According to Shannon–Hartley's theorem, the theoretical upper limit of the communication capacity C (bit/s) is

$$C = B \log_2 \left(1 + \frac{S}{N}\right). \tag{1.1}$$

where B (Hz) is the bandwidth, S (W) is the signal power, and N (W) is the noise power. Obviously, when B increases, the communication capacity C increases. Although the frequency of terahertz is overwhelmingly higher than that of microwaves, the frequency bandwidth also increases accordingly. This is shown in Figure 1.7. When information is modulated at a certain frequency, a finite frequency bandwidth is required. Let us suppose that the frequency bandwidth is $\pm 5\%$ of the center frequency. In the case of the center frequency of 3 GHz, the frequency bandwidth is 300 MHz from 2.85 to 3.15 GHz. On the other hand, at 300 GHz, the frequency bandwidth is 30 GHz from 285 to 315 GHz. That is, if the frequency bandwidth with respect to the center frequency (this is called the relative band) is constant, by increasing the center frequency, the frequency band automatically increases. This is a merit from the viewpoint of the user.

This advantage clearly appears in the growth of past data rates. Figure 1.8 shows the past growth of wired and wireless data rates. The data rate of wired communication is improving at a 10-fold pace in about 7.5 years. The data rate of wireless communication, on the other hand, has improved at a 10-fold pace in 4 years. That is, the evolution of the data rate of wireless communication is overwhelmingly greater. In both cases, improvement in the performance of semiconductors is based on the improvement of data rate. However, the contribution of semiconductor is somewhat different between wired and wireless. Wired communication uses a frequency band starting from DC called baseband. Of course, the higher the data rate, the wider the frequency band used. For this reason, the bandwidth is limited by the performance of the transistor. However, baseband circuits require good performance to low frequencies close to DC, and cannot be optimized for only high frequencies. Thus, the bandwidth is not only determined by the maximum operating frequency of the transistor. Therefore, in the case of wired communication, it is important to operate the circuit with as high frequency as possible from DC. On the other hand, in the case of wireless, the bandwidth at high frequencies mainly determines the data rate. Of course, as with wired communication, appropriate performance is required for baseband. But frequency-band restriction on high frequency is more severe than baseband since many users are already used and the available bandwidth at high frequencies is limited regardless of the technical constraints. Note that generally, as the frequency increases, the allocated frequency band becomes wider. Therefore, in wireless, raising the frequency is essentially important for improving the data rate. To raise the frequency, it is necessary to improve the maximum operation frequency of

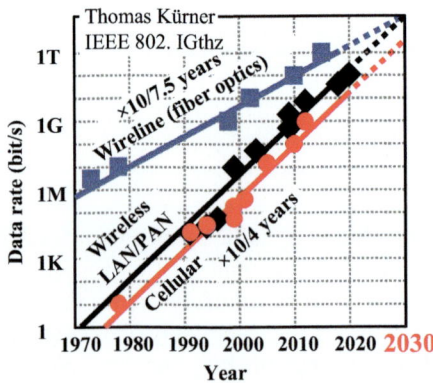

Figure 1.8   Evolution of data rates of wired and wireless communications. In 2020, wireless will reach 100 Gbit/s and could catch up with fiber-optic speeds in 2030

the transistor. Until a while ago, high-frequency performance improved dramatically due to the benefits of miniaturization of semiconductor process technologies. Due to this benefit, the operating frequency in wireless was improved and the data rate was improved. On the other hand, the signal-processing technology also improved due to large-scale integration of digital circuits, and the amount of digital information to be sent improved in the same bandwidth. Owing to both contributions, wireless has achieved a staggering evolution of ten times in 4 years.

On the other hand, in commercial cellular phones used in outdoor wireless, the generation is changing every 10 years. As shown in Figure 1.9, the maximum data rate was 100 Mbit/s in the third generation (3G), which started in 2000. Thereafter, the maximum data rate is 1 Gbit/s in the fourth generation (LTE), and the maximum data rate will be 10 Gbit/s in fifth generation (5G) started in 2020. If the alternation of cellular generation continues every 10 years, the sixth generation (6G) will appear in 2030. Considering that the maximum data rate is 10-fold for each generation, the maximum data rate of 6G reaches 100 Gbit/s. High frequencies are required, of course, to achieve this data rate, but frequencies have already been assigned to various uses. Frequency allocation is determined in each country, and Figure 1.10 is an allocation table of the Federal Communications Commission (FCC) in the United States. Frequencies are similarly allocated worldwide, and there is no freely allocated frequency below 275 GHz. To realize 100 Gbit/s, it is necessary to positively utilize over 275 GHz. Figure 1.11 shows the allocation of frequencies around 300 GHz, including 275 GHz. This is the background of the 300 GHz band wireless communication that this book is targeting.

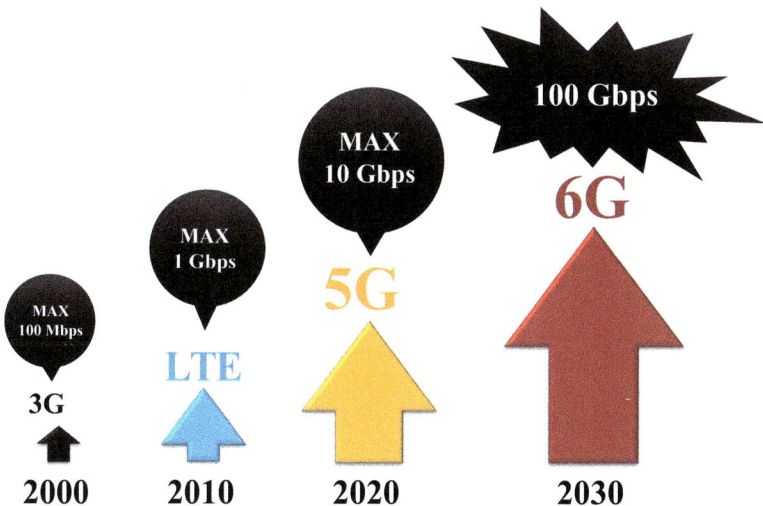

Figure 1.9   Cellular phones evolve every 10 years

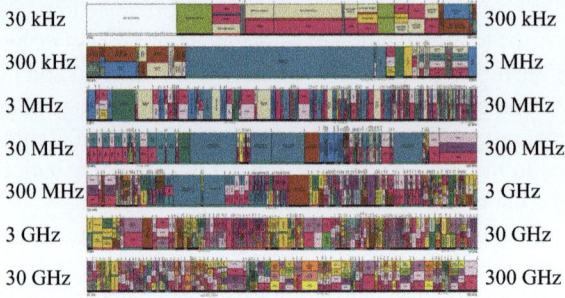

Figure 1.10   *Frequency allocation table in the United States. Frequency band below 275 GHz is fully allocated*

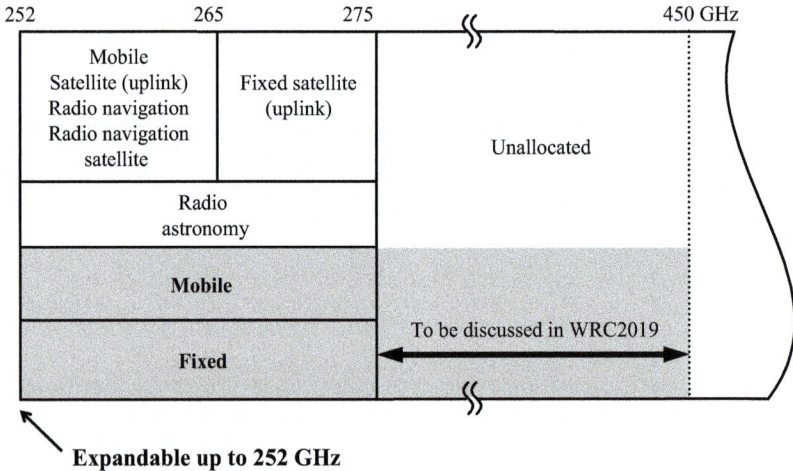

Figure 1.11   *Target frequency range for 300 GHz band wireless communication. Copyright 2017 IEICE. Reproduced, with permission, from [2]*

1.2   300-GHz-band wireless that realizes 100 Gbit/s

There are no frequencies that can be freely assigned below 275 GHz. On the other hand, discussion on frequency allocation exceeding 275 GHz is under way. As shown in Figure 1.12, the discussion on frequency allocation is progressing so that frequency range from 275 to 450 GHz can be used for communication as well. On the other hand, frequency range from 252 to 275 GHz has already been allocated for communication use. That is, frequency bands from 252 to 450 GHz will be continuously used for communication. This is the whole image of the target frequency of the 300 GHz band wireless.

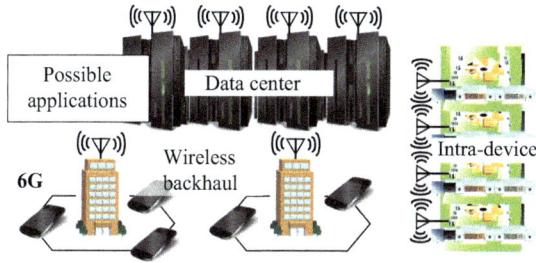

Figure 1.12   Potential applications for 300 GHz band wireless communication

What is the application of terahertz communication including this 300-GHz band? There are several groups discussing possible applications for terahertz communication. An example of potential applications is reported by the IEEE 802.15 Terahertz Interest Group. As shown in Figure 1.12, backhaul wireless communication, wireless data center and, intra-device communication are considered candidates. These are short-distance and fixed communications, and partial replacement of the current fiber-optic communication with wireless communication. Although these applications for terahertz communication are easy to imagine, it will be difficult to become popular just by replacing fiber-optic communication. Will the other applications than replacement of the existing fiber-optic communication be impossible? Will it be usable for mobile applications that broaden the application of wireless communication? Before discussing new applications, we would like to discuss why terahertz communication is limited for short-distance communication.

The first reason why applications for terahertz communication are limited in short range is atmospheric attenuation. Figure 1.13 shows the frequency characteristics of atmospheric attenuation. This figure includes a millimeter-wave band from 30 to 300 GHz and shows atmospheric attenuation from 30 GHz to 3 THz. Overall, the atmospheric attenuation has consistently increased with increasing frequency. Kilometer-class communication will be difficult above the line of 10 dB/km. The corresponding frequencies are 60 GHz due to oxygen absorption, 183 and 325 GHz due to water absorption, and all frequencies above 360 GHz. Long-distance communication is difficult with these frequencies. On the other hand, at frequencies other than these, the possibility of kilometer-class communication remains. That is, by selecting the frequency appropriately, kilometer-class communication can be possible even for terahertz, theoretically.

Figure 1.14 shows frequency characteristics of the distance at which atmospheric attenuation becomes 10 dB. As described above, the frequency range of communication in the 300-GHz band is from 252 to 450 GHz. It should be noted here that from 252 to 292 GHz, the distance at which the atmospheric attenuation is 10 dB exceeds 3 km. Communication distance will be shortened when it is rainy since this distance is calculated based on atmospheric attenuation without raining. But it is still worth noting that there is a possibility of kilometer-class communication.

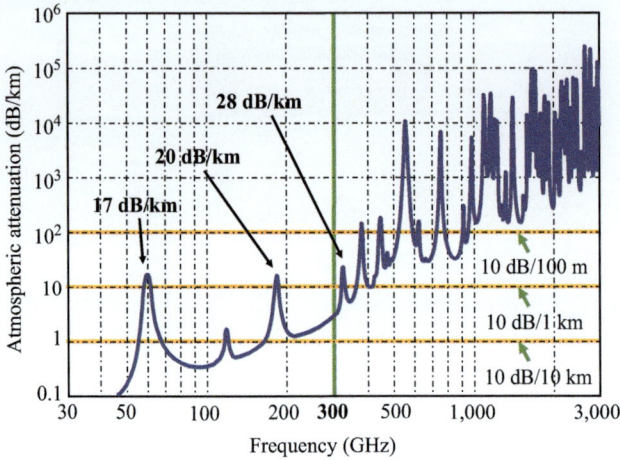

Figure 1.13   Atmospheric attenuation in millimeter waves and submillimeter waves (terahertz)

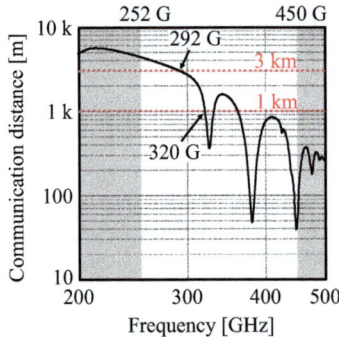

Figure 1.14   Frequency characteristics of communication distance. The distance, at which the atmospheric attenuation is 10 dB, is defined as the communication distance

In Figure 1.15, the frequency range of the 300 GHz band and the target communication distance are summarized. A frequency band of 40 GHz from 252 to 292 GHz can be used for medium-to-long-distance communication up to 3 km. With this frequency band, assuming 16 QAM (quadrature amplitude modulation) with the roll-off coefficient of 0.25, a data rate of 128 Gbit/s is realized. Also, if the communication distance is 1 km, the frequency range can be expanded up to 320 GHz. Now, the frequency band is 68 GHz, and the data rate of 218 Gbit/s is realized with 16 QAM. Furthermore, if it is a short distance communication within 100 m, it can be utilized up to 450 GHz, the data rate can theoretically be increased to 640 Gbit/s, and it approaches the terabit communication.

Frequency (GHz)

252   292   320   450

Medium-to-long distance (<3 km)
BW: 40 GHz / DR: 128 Gbps

Assumed 16 QAM

Medium distance (<1 km)
BW: 68 GHz / DR: 218 Gbps

Short distance (<100 m)
BW: 200 GHz / DR: 640 Gbps

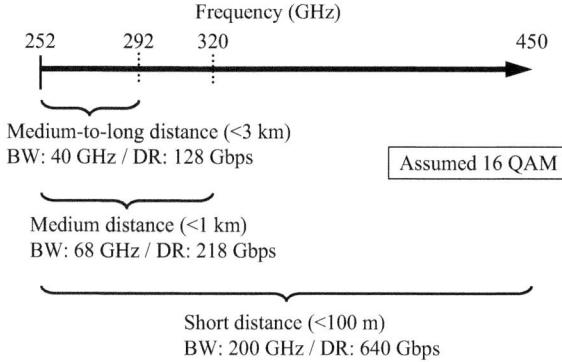

Figure 1.15   *Summary of relationship between frequency range, communication*
distance, and data rate in 300-GHz-band wireless communication

Another obstacle to the long-range communication by terahertz is propagation loss. According to Friis' transmission equation,

$$\frac{P_r}{P_t} = G_t G_r \left(\frac{\lambda}{4\pi d}\right)^2. \tag{1.2}$$

where P_r and P_t are the received and transmitted powers, respectively, G_t and G_r are gains of transmitting and receiving antennas, respectively, λ is the wavelength, and d is the communication distance. P_r/P_t is the propagation loss. According to this formula, the received power decreases in proportion to the square of the wavelength. That is, as the frequency increases, the propagation loss increases in proportion to the square of the frequency. On the other hand, the antenna gain G_I is given by

$$G_I = \frac{4\pi}{\lambda^2} A_e, \tag{1.3}$$

where A_e is the effective antenna area. The antenna gain is not only proportional to the effective antenna area but also increases in inverse proportion to the square of the wavelength. When this formula is used to replace Friis' transmission equation with the effective antenna area, we obtain

$$\frac{P_r}{P_t} = A_t A_r \left(\frac{1}{\lambda d}\right)^2. \tag{1.4}$$

According to this equation, the received power increases in inverse proportion to the square of the wavelength, and the propagation loss decreases in proportion to the square of the frequency. This shows the opposite relation to the relationship between wavelength and frequency and received power and propagation loss in the original Friis' transmission equation. In (1.2), the antenna gains are coefficients, and in (1.4), the effective antenna areas are coefficients. That is, if the antenna gain is constant, the propagation loss increases as the frequency increases, but if the effective antenna area is constant, the propagation loss decreases with increasing frequency. Of course, if the antenna gain is constant, the antenna can be miniaturized as the frequency increases. However, if the antenna size is reduced, the received power naturally becomes small. To avoid an increase in the propagation loss due to a constant antenna gain, even if the frequency becomes high, the antenna should not be miniaturized. The miniaturization of the antenna is one of the reasons for hindering long-distance communication.

The effective area A_I of the isotropic antenna is given by

$$A_I = \frac{\lambda^2}{4\pi}. \tag{1.5}$$

Some examples of the effective area of this isotropic antenna are shown in Figure 1.16. The effective antenna area is 7,262 mm² at 1 GHz, but only 0.08 mm² at 300 GHz. In this way, if the gain is constant, the area of the antenna is very small at 300 GHz, and the receiving antenna can capture only a small amount of power propagating through the space. On the other hand, if an antenna with the same effective area as 1 GHz is used at 300 GHz, the gain will rise to 50 dB i. As a result, it is necessary to use a high gain antenna in the 300 GHz band.

Consider the case study shown in Figure 1.17. In the 60 GHz band, the frequency band per channel is 2.16 GHz. Suppose the antenna gain is 14 dB i for both a transmitter and a receiver using a typical small patch antenna. What kind of antenna should be used in the 300-GHz band? Here, we make two assumptions. The first assumption is that the transmission power is constant. Of course, making the transmitting power in the 300 GHz band equal to that in the 60 GHz band requires a big technical challenge, but we will not consider it here. As a second assumption, atmospheric attenuation is neglected as a short-range application. In the first case, we will use the same frequency band 2.16 GHz as the 60-GHz band even in the 300-GHz band. Since the frequency band does not change from the 60-GHz band, in this case, the data rate does not improve. However, according to (1.2), to obtain the same received power, it is necessary to increase the antenna gain of transmitter and receiver up to 21 dB i. In the second case, to make the communication speed four times, the frequency band is assumed 8.64 GHz. Now, the signal to noise ratio (SNR) deteriorates in the case of

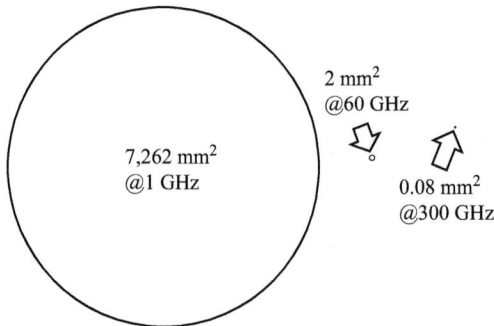

Figure 1.16   Schematic diagram of comparing the effective area of an isotropic antenna

	TX	RX	
60 GHz 2.16 GHz BW	◁ 14 dB i	▷ 14 dB i	How to realize the same communication distance?

Assumption 1: Output power is identical. (In practice, the same output power as 60 GHz band requires technical challenge.)

Assumption 2: Atmospheric attenuation is ignored.

	TX	TX
Case 1: 2.16 GHz bandwidth	◁ **21 dB i**	▷ **21 dB i**

	TX	TX
Case 2: 8.64 GHz bandwidth	◁ **24 dB i**	▷ **24 dB i**

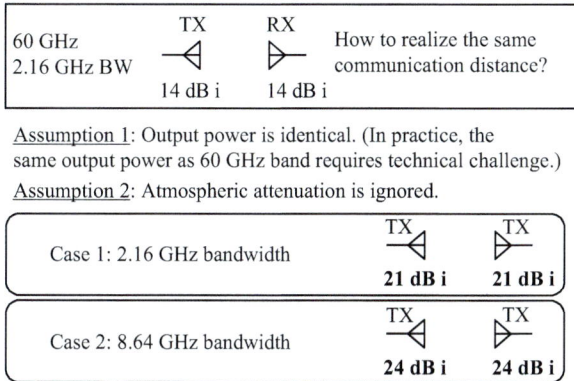

Figure 1.17   *Calculate the antenna gain of 300 GHz communication when communicating at a data rate equal to or higher than that in 60 GHz communication*

Figure 1.18   *Comparison of 300 and 60-GHz standard-gain horn antennas. Japanese 100-yen coins are placed, for reference*

the same received power. To keep the SNR constant, it is necessary to increase the antenna gain to 24 dB i. In either case, the antenna gain must be increased from 7 to 10 dB by changing from the 60 to 300 GHz band. The antenna gain is proportional to the effective antenna area, but what is the area of the high gain antenna in the 300 GHz band compared to the 60 GHz band?

Figure 1.18 shows photographs of standard gain horn antennas of 300-GHz band (WR 3.4) and 60-GHz band (WR 15). To understand the size, Japanese 100-yen coins are placed under the 300-GHz-band antenna. Since both are standard-gain horn antennas, the antenna gains are both 24 dB i. However, the size of the antenna in the 300-GHz band is very small, and even the long side is about 1 cm. Therefore, even with a high gain antenna, the antenna size does not become large. Here, higher gain narrows half-power beamwidth (HPBW). For example, the HPBW is 7.2° at 24 dB i and only 0.36° at 50 dB i. In terahertz communication, it is necessary to consider applications that take this high directivity into account. It is the reason why fixed applications are suitable for terahertz. However, even in mobile communication, if the relative angular velocity changes slowly, it will be possible to follow the direction of the antenna electronically or mechanically. Even if the relative speeds of the two transceivers are constant, the relative angular velocity decreases if they are far apart. Therefore, it is better to consider communication between objects far away if terahertz mobile communication is considered. This will be mentioned later.

It is also important, on the other hand, to consider the meaning of high speed in communication, for finding appropriate applications of terahertz communication. Normally, high-speed communication often refers to a high data rate. For example, data rates of (wired) fiber-optic communications are higher than wireless communications in general. However, low latency is one of the high-speed features. It can be said that low latency is real time. Since radio waves propagate through the atmosphere in wireless communication, its speed is almost equal to the speed of light in vacuum. On the other hand, in the case of fiber-optic communication, light propagates through the fiber, so that the speed is slowed down by the dielectric constant of the fiber. In general, the propagation speed on the radio is 50% higher than that of the optical fiber as shown in Figure 1.19. In other words, wireless communication is more excellent in real time property than fiber-optic communication. This "minimum latency" is crucial for applications requiring real-time responses over a long distance, including high-frequency trading [1]. As stock trading places importance on real-time nature, from 2011 to 2013, $250 million was invested to establish a wireless link between New York and Chicago.

Sometimes, the real-time nature is important for image transmission. For example, if a data rate exceeding 100 Gbit/s in the 300 GHz band is used, it is possible to transmit an uncompressed 8K image in real time. When 252–292 GHz is used, a communication distance of 3 km can be expected, real-time 8K image transmission at medium-to-long distances can be realized. For example, as shown in Figure 1.20, if an 8K camera is mounted on the drone, ultrahigh-definition images at a remote place can be transmitted to the base station in real time. If relay drones are also used, it will

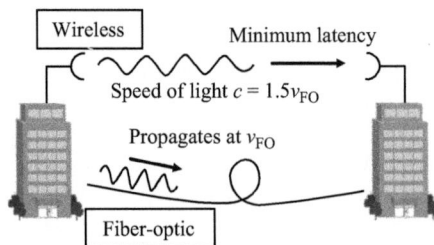

Figure 1.19   Comparison of the speed of wireless and fiber-optic communications in terms of latency

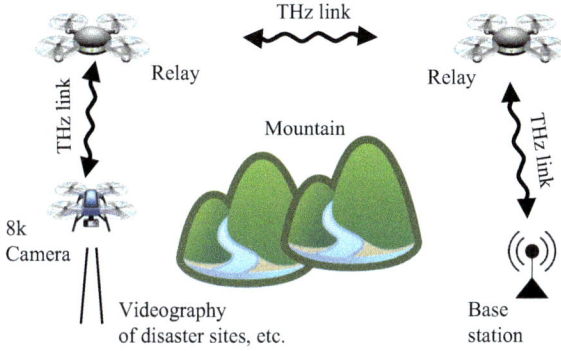

Figure 1.20   Real-time transmission of 8K video images using drones

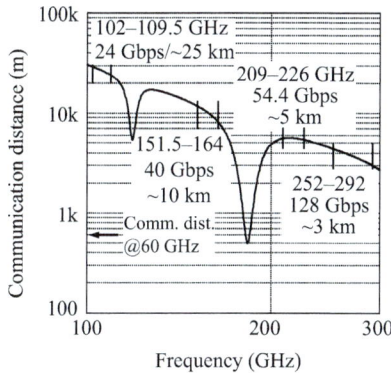

Figure 1.21   Relationship between frequency allocated for mobile communication and communication distance

be possible to transmit out of line of sight or a distance exceeding 3 km. Such applications are useful for disaster relief and search for victims. It will also help to monitor the site before an ambulance or a police car arrives at a traffic accident site. Because wired communication cannot be used for flight objects like drones, the advantage of wireless communication can be demonstrated. Besides, real-time transmission of 8K video can be used for sports broadcasting used in places such as golf courses and athletics venues. If the distance is in the kilometer range, the problem of strong directivity will be alleviated since the relative angular velocity during movement will also be small.

The frequencies used for such mobile communication is not limited from 252 to 292 GHz. Figure 1.21 shows the relationship between the frequency allocated as mobile communication and the communication distance. Four frequency bands from 100 to 300 GHz have been allocated for mobile communication. As the frequency decreases, the communication distance increases while the bandwidth narrows. Here, at 60 GHz, which is nearing practical use as high-speed communication, the atmospheric attenuation is large and the communication distance is only several hundred meters. In contrast, the terahertz communication can communicate with a much longer distance. Therefore, if the frequency is used properly according to the application, it is expected that new applications will be developed for mobile communication.

1.3   Recent progress of terahertz integrated circuits for communications [2]

As described in Section 1.2, frequencies around 300 GHz offer extremely broad atmospheric transmission window with relatively low losses of up to 10 dB/km and can be regarded as the ultimate platform for ultrahigh-speed wireless communications with near-fiber-optic data rates. In this section, we will review technical challenges and recent advances in integrated circuits (IC) targeted at communications using these and nearby "terahertz (THz)" frequencies.

1.3.1   Review of 300-GHz band wireless communication

The 300-GHz band has fairly low atmospheric losses of roughly below 10 dB/km and potentially allows near-fiber-optic wireless data rates. In theory, as wide as 68 GHz bandwidth (252–320 GHz) is available for allocation for wireless communications. It is the only frequency band that has such acceptable losses and such a wide bandwidth. The 300-GHz band is arguably the last and the ultimate frequency band for ultrahigh-speed communications. We therefore also discuss possible channel allocation plans for the 300-GHz band considering compatibility with existing wireless standards in Section 5.1. Photonics-based approaches to THz wireless communications [3,4] preceded IC-based approaches. However, research into THz wireless ICs is becoming very active and even THz CMOS ICs have been reported. In this section, we will review and discuss recent THz ICs for wireless communications and their performance. We will, therefore, not cover other approaches, including those based on resonant-tunneling diodes [5].

1.3.2   300-GHz transmitters and receivers

There are different ways of generating 300-GHz-band EM waves. In photomixing technology [3,4], two light sources having different wavelengths are used to generate a beat note at a THz frequency. Optical-to-electrical conversion is typically done using a uni-traveling-carrier photodiode (UTC-PD) [6].On the other hand, typical IC-based THz generation involves frequency multipliers. Recent InP [7] and GaAs [8] technologies have a transistor unity-power-gain frequency, f_{max}, higher than 1 THz. SiGe bipolar technology with $f_{max} \gtrsim 500$ GHz has also been reported [9]. Amplifiers operating in the 300-GHz band can be built using a technology with $f_{max} \gtrsim 500$ GHz.

When using such a high-f_{max} technology, a transmitter architecture with a power amplifier (PA) can be adopted. Kim *et al.* [10] reported a 300-GHz transmitter with the architecture shown in Figure 1.22(a). Sarmah *et al.* [11] showed a 240-GHz SiGe transmitter and Kallfass *et al.* [12] showed a 300-GHz GaAs transmitter using the architecture with a quadrature mixer, shown in Figure 1.22(b). The architectures shown in Figure 1.22 are essentially the same as those used by lower frequency transmitters.

There were also some reports on above-f_{max} circuits using moderate-f_{max} technologies, typically CMOS or BiCMOS. In this case, it is not possible to build THz-band amplifiers. Figure 1.23 shows architectures with a frequency multiplier at the final stage. Since multipliers are fundamentally nonlinear, a modulated signal undergoes distortion when fed into a multiplier. Hu *et al.* [13] reported a SiGe on-off keying (OOK) transmitter at 400 GHz using the architecture shown in Figure 1.23(a). OOK is not affected by the multiplier. Certain combinations of modulation format and multiplication factor actually allow the use of phase-shift keying (PSK). For example, Kang *et al.* [14] adopted the architecture shown in Figure 1.23(b) and demonstrated a frequency-tripler-last 240-GHz CMOS transmitter that supports quaternary PSK (QPSK). Symbol permutation takes place when a QPSK-modulated signal is fed into a tripler, but the resulting signal constellation is still that of QPSK and demodulation is possible. In general, when the modulation is M-PSK and the frequency multiplication factor is m, signal constellation is retained (though with symbol permutation) if m is relatively prime to M. In [14], $M = 4$ and $m = 3$. A drawback of this architecture is that the signal bandwidth is also multiplied by m, and spectral efficiency does not improve much even with $M > 2$.

To support QAM and/or to prevent signal bandwidth spreading, an upconversion mixer is needed. Park *et al.* [15] presented a 260-GHz CMOS OOK transmitter using the architecture shown in Figure 1.24(a). Lopez-Diaz *et al.* [16] presented a

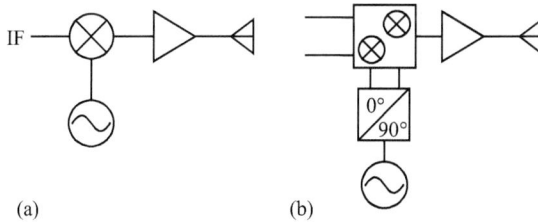

Figure 1.22   *Power-amplifier (PA)-last THz transmitter architectures with (a) intermediate-frequency (IF) input and (b) baseband (BB) input. Copyright 2017 IEICE. Reproduced, with permission, from [2]*

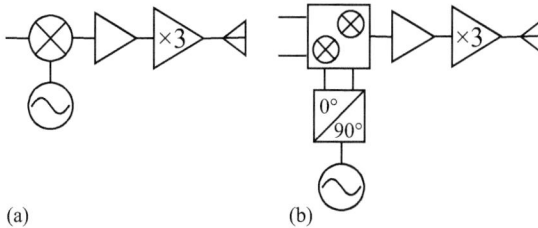

Figure 1.23   *Frequency-multiplier-last THz transmitter architectures with (a) intermediate-frequency (IF) input and (b) baseband (BB) input. Copyright 2017 IEICE. Reproduced, with permission, from [2]*

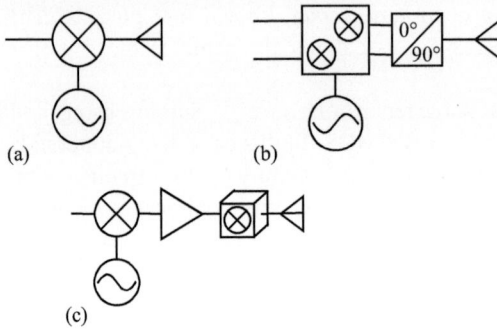

Figure 1.24   *Mixer-last THz transmitter architectures with (a and c)*
intermediate-frequency (IF) input and (b) baseband (BB) input.
(c) Cubic or square mixer is used for the last stage. Copyright 2017
IEICE. Reproduced, with permission, from [2]

240-GHz GaAs transmitter with a quadrature mixer using the architecture shown in Figure 1.24(b). These architectures use a mixer that has two inputs at the last stage, and therefore it is difficult to parallelize the intermediate frequency (IF) stage and perform power combining after upconversion. At THz frequencies, the output power obtained from a single mixer is very low especially when the transmitter is operating above f_{max}. Parallelization and power combining therefore is a must. To facilitate it, we used the architecture shown in Figure 1.24(c) [17–20]. The mixer in this architecture, called a *cubic* or *square mixer*, is a tripler- or doubler-based subharmonic mixer, respectively, and receives superposition of two signals, local oscillator (LO) and modulated IF. A THz signal is generated by the cubic nonlinearity of the tripler and the quadratic nonlinearity of the doubler. This is analogous to photomixing, in which superposition of two light waves with different wavelengths is fed into a nonlinear device (UTC-PD). This architecture does not cause signal bandwidth spreading and allows massive parallelization and power combining. In addition, since the cubic and square mixers have good linearity, this architecture supports QAM [17–20], which will be discussed in Section 4.1.

Figure 1.25 shows THz receiver architectures reported recently. Depending on the value of f_{max}, a low-noise amplifier (LNA) may or may not exist after the receiving antenna. The mixer configuration depends on the modulation format that the receiver supports. Overall, the architectures are simpler than those of transmitters. The architecture shown in Figure 1.25(a) was adopted by a 300-GHz InP receiver [10]. The architecture shown in Figure 1.25(b) was adopted by a 400-GHz SiGe receiver [13] and a 260-GHz CMOS receiver [15]. The architecture shown in Figure 1.25(c) was adopted by a 240-GHz SiGe receiver [11]. The architecture shown in Figure 1.25(d) was adopted by a 240-GHz CMOS receiver [21]. In the mixer-first architecture shown in Figure 1.25(b) and (d), the noise figure of the following stages is not suppressed by an LNA. To suppress deterioration of the noise figure, it is necessary to reduce conversion loss in the mixer. Therefore, we realized a 300-GHz band CMOS receiver

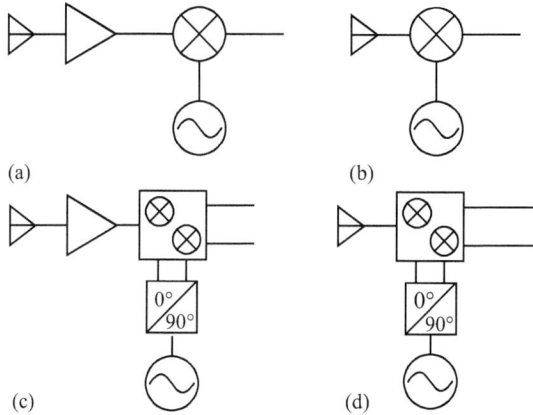

Figure 1.25   THz receiver architectures. (a) LNA-first/IF output, (b) mixer-first/IF output, (c) LNA-first/BB output, and (d) mixer-first/BB output. Copyright 2017 IEICE. Reproduced, with permission, from [2]

using fundamental-mixer-first architecture. The technique will be discussed in detail in Section 4.2.

1.4   Recent progress of terahertz CMOS circuits [23]

Now, in this section, recent progress of terahertz CMOS ICs is reviewed. The development of terahertz CMOS is somewhat different from the conventional digital and RF CMOS evolution, in which it is not fueled as much by technology scaling. Rather, the key enablers progress in high-frequency device characterization and modeling techniques and in design techniques near transistor's active operation limit, f_{max}.

1.4.1   Technical background of terahertz CMOS circuit design

CMOS 60-GHz Wi-Fi products are nearing mass production, and 5G cellular network is also another platform where millimeter-wave CMOS products are likely to play a vital role [22]. Development of 77- and 79-GHz CMOS radar front end has also been very active. For the abovementioned frequency bands, the target applications have long been clear, and the well-motivated development of CMOS millimeter-wave front end has generally been successful. This success owes very much to the device performance improvement by means of miniaturization. At frequencies above 100 GHz, target applications are not as obvious yet. Nevertheless, research into ICs covering these "terahertz" frequencies is becoming active these days. One important aspect of terahertz CMOS is that its development is not as tightly linked with CMOS technology advancement by miniaturization. The latter is an ongoing effort but is geared toward digital circuits, not extreme frequencies. State-of-the-art CMOS processes can even be a suboptimal choice for terahertz circuits for two reasons. One is because passive components cannot be made smaller by migrating to a more advanced process. While digital circuits can enjoy cost reduction due to device and interconnect geometry shrinking, terahertz circuits will inevitably include multistage amplifiers with many passive components that occupy a large die area, implying undue increase in the costs. The other is that the better current drivability of the MOSFET due to reduced channel length does not necessarily lead to commensurately higher unity-power-gain frequency, f_{max}. Here the power gain is either the maximum available gain or Mason's unilateral gain [24]. As a result, 65-nm CMOS is often used for terahertz circuits for offering reasonable balance between performance and cost. But if device performance improvement is not there to help, what is it that is enabling CMOS designers to tackle terahertz circuit development? The key enablers are recent progress in high-frequency device characterization and modeling and also in design techniques near transistor's active operation limit, f_{max}. In the following, we review these aspects of terahertz CMOS research in recent years.

1.4.2   Device characterization and modeling

1.4.2.1   De-embedding

Predictive circuit simulation is possible only if accurate device models are available. A device model is, at best, as accurate as the measurement data from which it is built. It is, therefore, very important that measurement data are reliable. Making measurements at high frequencies involves de-embedding of device characteristics from raw measurement data. At terahertz frequencies, not only does measurement uncertainty increase, also possible excitation of spurious EM modes can complicate the interpretation of measurement results. Many popular methods of de-embedding and models of pads used at low-GHz frequencies are inadequate at such high frequencies. Instead, the thru-reflect-line (TRL) algorithm is commonly used for characterizing transmission lines and for subsequent de-embedding. It was recently shown that the lengths of line standards used in TRL have a significant impact on the reliability of the extracted propagation constant, γ, of the line and de-embedding [25]. A typical longest line length that can be accommodated on a typical CMOS die of, say, 4 mm $\times$ 4 mm would be around 3 mm. The phase constant β can be extracted reliably from a line of that length, but the attenuation constant α can show erratic behavior above 200 GHz that is difficult to predict by EM simulation. On the other hand, if the longest line is as long as 8 mm, not only the phase constant but also the attenuation constant agree reasonably well with EM simulation up to 300 GHz. This is a recent finding that significantly improved reliability of transmission line characterization and device de-embedding [25]. It will be discussed in Section 3.2.2.

1.4.2.2   Electromagnetic field simulation

Terahertz circuit design cannot be done with circuit simulation alone. EM field simulation is a must for predictive design. The free-space EM wavelength at 100 GHz is only 3 mm. The on-chip wavelength of around 1.5 mm (or shorter at higher frequencies) cannot be neglected even on a small chip, and circuits must be treated as distributed. The common layout parasitic extraction used in RF circuit simulation ignores inductance and is inadequate for terahertz circuit simulation. EM field simulation is needed to design various passive millimeter-wave components as well as transmission lines. Just as predictive circuit simulation requires accurate device models, predictive EM field simulation requires accurate information about process parameters, including material parameters, layer thicknesses, and cross-sectional geometry of interconnects. CMOS foundries provide nominal values of most of these parameters. But permittivities of dielectrics they provide are usually frequency-independent real values. The assumption of frequency independence is not unreasonable, but the lack of information about dielectric loss is a problem. To perform EM simulation with reasonable predictive power, material parameters (dielectric permittivities and metal conductivities) must be evaluated by measurement. The measurement data that can be used for this purpose are S parameters measured with a vector network analyzer (VNA). Several dielectric materials are used in modern CMOS interconnect layers. Since the designer must follow the restrictive CMOS design rules, it is not possible to design test structures that reveal material parameters of each individual material. The test structures should be chosen appropriately so that measured characteristics could be related straightforwardly to material parameters. Transmission lines used in TRL are ideally suited to this purpose because the extracted propagation constant directly reflects effective material parameters. These are effective parameters because multiple materials are involved and actual geometrical parameters are not known exactly. By measuring a few different types of transmission line, effective material parameters can be estimated fairly reliably [26,27]. It will be discussed in Section 3.2.3.

1.4.2.3 Nonlinear transistor model

Device characterization at high frequencies is done primarily by measurement of scattering matrices or S parameters with a VNA. S parameters are linear responses of a device at the measurement frequency. Large-signal network analyzers that measure harmonic as well as fundamental responses are also available for low-GHz frequencies [28]. X-Parameters are one such proprietary framework for frequency-domain nonlinear measurement and simulation [29]. However, harmonic responses cannot currently be measured at terahertz frequencies. For one thing, the usable frequency ranges of terahertz waveguides are too narrow for harmonics to be handled. Typical measurement-based (as opposed to frequency-extrapolated) terahertz device models are small-signal models built from S parameters. But obviously, small-signal models are inadequate for simulating PAs, oscillators, and mixers. Predictive nonlinear circuit simulation is a challenge at lower frequencies too. But millimeter-wave circuits operating below 100 GHz can be realized by overdesign even if the predictive ability of simulation is poor. On the other hand, at terahertz frequencies, transistor's gain is so limited and passive components are so lossy that there is very little room for overdesign. There, therefore, is much greater demand for accurate large-signal models. Study of nonlinear microwave characterization and modeling of transistors has focused mainly on compound semiconductor devices and high-voltages MOSFETs such as LDMOS (laterally diffused MOS) FETs [30,31]. The primary focus of highly miniaturized MOSFET modeling, on the other hand, has been compact and scalable models for digital circuits. Rather different approaches were taken to the modeling of microwave transistors and digital MOSFETs. Considerable effort remains to be put forth at adapting microwave-transistor modeling techniques to MOSFET modeling. Parasitic resistances at gate, source and drain measured at DC are not necessarily equal to those extracted from high-frequency measurements. Such seeming discrepancy must be appropriately characterized and modeled for fully predictive circuit simulation at different time scales [32]. It will be discussed in Section 3.3.

1.4.3   Terahertz CMOS building-block design

1.4.3.1   Gain boosting in amplifiers

Terahertz circuits operate at frequencies that are much closer to MOSFET's f_{max} than usual. Typical RF and millimeter-wave circuits operate at below $f_{max}/4$ [33], whereas terahertz circuits often have to operate at around $f_{max}/2$ or higher. A common local feedback technique for gain boosting is to insert a positive reactance between the gate and drain of a MOSFET. This technique, known as neutralization, was originally proposed to cancel the unwanted internal feedback in a transistor and unilateralize the amplifier [34,35]. The positive reactance can be realized most easily in a differential amplifier with a cross-coupled capacitor pair because the cross-coupled capacitance is, in differential mode, equivalent to negative capacitance [36]. It is known that by adopting cross-coupled feedback capacitance larger than the gate-drain capacitance, the gain can become higher than the unilateral gain. Essentially, the same feedback can be achieved in single-ended amplifiers by transmission-line feedback [37]. That these apparently different feedback configurations are essentially the same can be seen within the framework of a general theory [38]. It will be discussed in Section 2.1.

1.4.3.2   High-frequency generation

Near-f_{max} or above-f_{max} frequency generation requires frequency multiplication. An N-stage ring oscillator can generate Nth harmonic. The case of $N = 2$ is known as the push–push oscillator [39], and its output power is relatively large. The larger the value of N, the smaller the output power of the Nth harmonic. In CMOS terahertz circuits, the use of $N = 3$ (triple-push oscillator) is common. Since the output power of a single oscillator is small, power-combining techniques are often applied. When outputs from multiple oscillators are to be combined in parallel, the output signals must be in phase. One way to accomplish the mutual synchronization is to use injection locking [40]. Power combining can also be done spatially by appropriately disposing multiple terahertz sources and antennas, thereby avoiding power loss in power-combining passive components [41]. A loop of coupled and mutually injection-locked oscillator was recently proposed. A low phase noise of -92 dB c/Hz at 1-MHz offset from a 100-GHz carrier frequency was reported [42]. As indicated in these examples, significant progress has been made in making output power higher and phase noise lower.

1.4.3.3   On-chip power-line decoupling, planar circuits and antennas

Power-line decoupling for millimeter-wave circuits is much more challenging than for RF circuits because of the much wider operation frequency range. The use of decoupling capacitors can be problematic due to their low self-resonance frequency. A better alternative is to use as power lines a specially designed transmission lines having extremely low characteristic impedance [43]. It was recently demonstrated that such a power line's input impedance does stay as low as 1 Ω up to 170 GHz and below 2 Ω up to 325 GHz [44,45]. It will be discussed in Section 3.1.2. In millimeter-wave CMOS circuits, baluns and other coupled- or uncoupled-transmission line-based components are used, as well as ordinary capacitors and inductors. In addition, as the frequency becomes higher, more extensive use is made of planar circuit components and techniques. For example, mixed use of slotlines and microstrips, a known technique in microwave printed circuit boards, is now made in CMOS ICs [46]. The use of hollow waveguides in ICs is also being investigated [47]. Research on such unconventional techniques is very active. On-chip antennas are also becoming increasingly common. CMOS on-chip antennas tend to be significantly lossier than off-chip antennas due to the presence of silicon substrate. However, it is difficult to realize low-loss wired connection to off-chip components at terahertz frequencies, and on-chip antennas are used to solve that problem [15,48].

1.4.4   Outlook

Throughout the history of CMOS RF circuit evolution, device-performance improvement due to technology scaling or device miniaturization contributed so much to circuit performance improvement that, arguably, the excellence of design has not played as important a role in shaping the long-term trend of continued performance growth. But now that it is becoming more and more difficult to enjoy the fruit of miniaturization as far as raising the operation frequency is concerned, design techniques built upon firm circuit theoretic, EM, or algorithmic foundations are assuming greater importance. The recent efforts at realizing terahertz CMOS circuits near f_{max}, reviewed above, can be seen as a reflection of the qualitative shift in technological trend: a departure from device-performance-powered design. And that departure will contribute to keep the technological trend at a higher level, the continued exponential growth (Moore's law and its direct and indirect corollaries). The demand for possible main terahertz applications of ultrahigh-speed communication, sensing, and imaging, which require high-performance digital blocks, could grow to a point where mass production with a state-of-the-art CMOS process is desirable. Even if so, given the fact that the growth of CMOS f_{max} is slowing down, the development of near-f_{max} design technology will remain important in the foreseeable future. CMOS millimeter-wave circuits have been developed based mainly on the traditional lumped-circuit design approach with some distributed or EM corrections. But there are a wealth of design concepts developed by microwave and antenna engineers, still unexplored yet broadly applicable to CMOS terahertz design. Collaborative efforts by CMOS designers and microwave and antenna engineers will most certainly accelerate the development and unearthing of full potential and impacts of terahertz CMOS.

References

[1]   C. Cookson, "Time is money when it comes to microwaves," *FT Magazine*, May 10, 2013.

[2]   M. Fujishima, S. Amakawa, "Integrated-circuit approaches to THz communications: Challenges, advances, and future prospects," *IEICE Transactions on Fundamentals of Electronics, Communications and Computer Sciences*, vol. 100, no. 2, pp. 516–523, 2017.

[3]   G. Ducournau, P. Szriftgiser, F. Pavanello, *et al.*, "THz communications using photonics and electronic devices: the race to data-rate," *Journal of Infrared, Millimeter, and Terahertz Waves*, vol. 36, no. 2, pp. 198–220, 2015.

[4]   T. Nagatsuma, G. Carpintero, "Recent progress and future prospect of photonics-enabled terahertz communications research," *IEICE Transactions on Electronics*, vol. E98-C, no. 12, pp. 1060–1070, 2015.

[5]   N. Oshima, K. Hashimoto, S. Suzuki, M. Asada, "Wireless data transmission of 34 Gbit/s at a 500-GHz range using resonant-tunnelling-diode terahertz oscillator," *Electronics Letters*, vol. 52, no. 22, pp. 1897–1898, 2016.

[6]   T. Ishibashi, S. Kodama, N. Shimizu, T. Furuta, "High-speed response of uni-traveling-carrier photodiodes," *Japanese Journal of Applied Physics*, vol. 36, no. 10, pp. 6263–6268, 1997.

[7]   W. R. Deal, A. Zamora, K. Leong, *et al.*, "A 670 GHz low noise amplifier with < 10 dB packaged noise figure," *IEEE Microwave and Wireless Components Letters*, vol. 26, no. 10, pp. 837–839, 2016.

[8]   I. Kallfass, F. Boes, T. Messinger, *et al.*, "64 Gbit/s transmission over 850 m fixed wireless link at 240 GHz carrier frequency," *Journal of Infrared, Millimeter, and Terahertz Waves*, vol. 36, no. 2, pp. 221–233, 2015.

[9]   A. Fox, B. Heinemann, H. Rücker, *et al.*, "Advanced heterojunction bipolar transistor for half-THz SiGe BiCMOS technology," *IEEE Electron Device Letters*, vol. 36, no. 7, pp. 642–644, 2015.

[10] S. Kim, J. Yun, D. Yoon, *et al.*, "300 GHz integrated heterodyne receiver and transmitter with on-chip fundamental local oscillator and mixers," *IEEE Transactions on Terahertz Science and Technology*, vol. 5, no. 1, pp. 92–101, 2015.

[11] N. Sarmah, J. Grzyb, K. Statnikov, *et al.*, "A fully integrated 240-GHz direct-conversion quadrature transmitter and receiver chipset in SiGe technology," *IEEE Transactions on Microwave Theory and Techniques*, vol. 64, no. 2, pp. 562–574, 2016.

[12] I. Kallfass, I. Dan, S. Rey, *et al.*, "Towards MMIC-based 300 GHz indoor wireless communication systems," *IEICE Transactions on Electronics*, vol. E98-C, no. 12, pp. 1081–1090, 2015.

[13] S. Hu, Y.-Z. Xiong, B. Zhang, *et al.*, "A SiGe BiCMOS transmitter/receiver chipset with on-chip SIW antennas for terahertz applications," *The IEEE Journal of Solid-State Circuits*, vol. 47, no. 11, pp. 2654–2664, 2012.

[14] S. Kang, S. V. Thyagarajan, A. M. Niknejad, "A 240 GHz fully integrated wideband QPSK transmitter in 65 nm CMOS," *The IEEE Journal of Solid-State Circuits*, vol. 50, no. 10, pp. 2256–2267, 2015.

[15] J.-D. Park, S. Kang, S. V. Thyagarajan, E. Alon, A. M. Niknejad, "A 260 GHz fully integrated CMOS transceiver for wireless chip-to-chip communication," *Symp. VLSI Circuits*, pp. 48–49, 2012.

[16]　D. Lopez-Diaz, I. Kallfass, A. Tessmann, *et al.*, "A subharmonic chipset for gigabit communication around 240 GHz," *IEEE MTT-S International Microwave Symposium*, pp. 1–3, 2012.

[17]　K. Katayama, K. Takano, S. Amakawa, *et al.*, "A 300 GHz 40 nm CMOS transmitter with 32-QAM 17.5 Gb/s/ch capability over 6 channels," *International Solid-State Circuits Conference*, pp. 342–343, 2016.

[18]　K. Katayama, K. Takano, S. Amakawa, *et al.*, "A 300 GHz CMOS transmitter with 32-QAM 17.5 Gb/s/ch capability over six channels," *The IEEE Journal of Solid-State Circuits*, vol. 51, no. 12, pp. 3037–3048, 2016.

[19]　K. Takano, K. Katayama, S. Amakawa, T. Yoshida, M. Fujishima, "A 300-GHz 64-QAM CMOS transmitter with 21-Gb/s maximum per-channel data rate," *European Microwave Integrated Circuits Conf.*, pp. 193–196, 2016.

[20]　K. Takano, K. Katayama, S. Amakawa, T. Yoshida, M. Fujishima, "Wireless digital data transmission from a 300 GHz CMOS transmitter," *Electronics Letters*, vol. 52, no. 15, pp. 1353–1355, 2016.

[21]　S. V. Thyagarajan, S. Kang, A. M. Niknejad, "A 240 GHz fully integrated wideband QPSK receiver in 65 nm CMOS," *The IEEE Journal of Solid-State Circuits*, vol. 50, no. 10, pp. 2268–2280, 2015.

[22]　K. Sakaguchi, G. K. Tran, H. Shimodaira, *et al.*, "Millimeter-wave evolution for 5G cellular networks," *IEICE Transactions on Communications*, vol. E98-B, no. 3, pp. 388–402, 2015.

[23]　M. Fujishima and S. Amakawa, "Recent progress and prospects of terahertz CMOS," *IEICE Electronics Express*, vol. 12, no. 13, 20152006, 2015.

[24]　M. S. Gupta, "Power gain in feedback amplifiers, a classic revisited," *IEEE Transactions on Microwave Theory and Techniques*, vol. 40, no. 5, pp. 864–879, 1992.

[25]　S. Amakawa, A. Orii, K. Katayama, *et al.*, "Design of well-behaved low-loss millimetre-wave CMOS transmission lines," *IEEE Workshop on Signal and Power Integrity*, pp. 1–4, 2014.

[26]　S. Amakawa, A. Orii, K. Katayama, *et al.*, "Process parameter calibration for millimeter-wave CMOS back-end device design with electromagnetic field analysis," *International Conference on Microelectronic Test Structures*, pp. 182–187, 2014.

[27]　K. Takano, K. Katayama, S. Mizukusa, S. Amakawa, T. Yoshida, M. Fujishima, "Systematic calibration procedure of process parameters for electromagnetic field analysis of millimeter-wave CMOS devices," *International Conference on Microelectronic Test Structures*, pp. 230–234, 2015.

[28]　P. Roblin, *Nonlinear RF Circuits and Nonlinear Vector Network Analyzers: Interactive Measurement and Design Techniques*, Cambridge University Press, Singapore, 2011.

[29]　D. E. Root, J. Verspecht, J. Horn, M. Marcu, *X-Parameters: Characterization, Modeling, and Design of Nonlinear RF and Microwave Components*, Cambridge University Press, Singapore, 2013.

[30]　P. H. Aaen, J. A. Plá, J. Wood, *Modeling and Characterization of RF and Microwave Power FETs*, Cambridge University Press, Singapore, 2007.

[31]   M. Rudolph, C. Farge, D. E. Root, editors, *Nonlinear Transistor Model Parameter Extraction Techniques*, Cambridge University Press, Singapore, 2012.

[32]   K. Katayama, S. Amakawa, K. Takano, M. Fujishima, "300-GHz MOSFET model extracted by an accurate cold-bias de-embedding technique," *IEEE MTT-S International Microwave Symposium*, pp. 1–4, 2015.

[33]   S. Voinigescu, *High-Frequency Integrated Circuits*, Cambridge University Press, Singapore, 2013.

[34]   C. C. Cheng, "Neutralization and unilateralization," *IRE Transactions on Circuit Theory*, vol. 2, no. 2, p. 138, 1955.

[35]   A. P. Stern, C. A. Aldridge, W. F. Chow, "Internal Feedback and Neutralization of Transistor Amplifiers," *IRE Transactions on Circuit Theory*, vol. 43, no. 7, p. 838, 1955. See https://ieeexplore.ieee.org/document/6373416

[36]   J. Choma, W. K. Chen, *Feedback Networks: Theory and Circuit Applications*, World Scientific, Singapore, 2007.

[37]   O. Momeni, *International Solid-State Circuits Conference*, p. 140, 2013.

[38]   S. Amakawa, *Asia-Pacific Microwave Conference*, p. 1184, 2014.

[39]   D. A. Boyd, *IEEE MTT-S International Microwave Symposium*, p. 587, 1987.

[40]   Y. M. Tousi, O. Momeni, E. Afshari, *International Solid-State Circuits Conference*, p. 258, 2012.

[41]   R. Han, E. Afshari, *International Solid-State Circuits Conference*, p. 138, 2013.

[42]   M. Adnan, E. Afshari, "A 105-GHz VCO With 9.5% Tuning Range and 2.8-mW Peak Output Power in a 65-nm Bulk CMOS Process," *IEEE Transactions on Microwave Theory and Techniques*, vol. 62, p. 753, 2014.

[43]   Y. Manzawa, M. Sasaki, M. Fujishima, "High-Attenuation Power Line for Wideband Decoupling," *IEICE Transactions on Electronics*, vol. E92-C, p. 792, 2009.

[44]   R. Goda, S. Amakawa, K. Katayama, K. Takano, T. Yoshida, M. Fujishima, *International Conference on Microelectronic Test Structures*, p. 220, 2015.

[45]   S. Amakawa, R. Goda, K. Katayama, K. Takano, T. Yoshida, M. Fujishima, *IEEE MTT-S International Microwave Symposium*, 2015.

[46]   R. Han, C. Jiang, A. Mostajeran, *et al.*, *International Solid-State Circuits Conference*, p. 446, 2015.

[47]   Y. Shang, H. Yu, H. Fu, W. M. Lim, "A 239–281 GHz CMOS Receiver With On-Chip Circular-Polarized Substrate Integrated Waveguide Antenna for Sub-Terahertz Imaging," *IEEE Transactions on Terahertz Science and Technology*, vol. 4, p. 686, 2014.

[48]   Z. Wang, P.-Y. Chiang, P. Nazari, C.-C. Wang, Z. Chen, P. Heydari, *International Solid-State Circuits Conference*, p. 136, 2013.

Chapter 2
Amplifier design

2.1 Amplifier theory

2.1.1 Introduction

Theoretical interest in small-signal amplifiers has traditionally centered around the unconditionally stable condition [1–4], where simultaneous conjugate matching is possible with passive source and load impedances, $Z_S = R_S + jX_S$ and $Z_L = R_L + jX_L$ (Figure 2.1), with $R_S > 0$ and $R_L > 0$. The theoretical upper bound gain is known as the maximum available gain (MAG). However, radio-frequency (RF) circuits in production are typically built with transistors having a highest f_{max} (maximum operation frequency) of four to five times the carrier frequency f [5]. Transistors are almost never unconditionally stable at such a frequency $f \lesssim f_{max}/4$, and a well-accepted rule of thumb in small-signal amplifier design is to set the gain well below the so-called maximum stable gain (MSG) of the transistor [6,7]. The MSG is given by $G_{ms} = |S_{21}/S_{12}| = |Z_{21}/Z_{12}|$ [6–9] and is also known as the figure-of-merit gain [8], for it is not the theoretical upper bound of the stable gain. Under conditionally stable conditions, the stability factor K [8,9] satisfies $-1 < K < 1$. Unconditional stabilization by introduction of some losses or negative feedback may also be performed, leading to $K > 1$ and a lower gain.

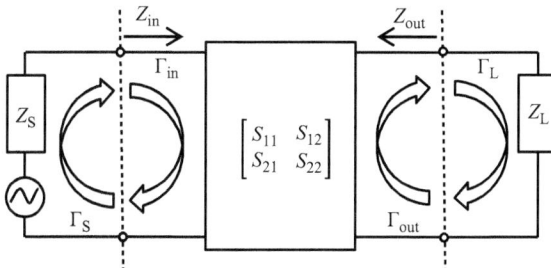

Figure 2.1 Linear two-port amplifier with a source impedance
$Z_S = R_{ref}(1 + \Gamma_S)(1 - \Gamma_S)$ and a load impedance
$Z_L = R_{ref}(1 + \Gamma_L)(1 - \Gamma_L)$, where the reference resistance is typically
$R_{ref} = 50\ \Omega$. The input and output reflection coefficients are
$\Gamma_{in} = (Z_{in} - R_{ref})/(Z_{in} + R_{ref})$ and $\Gamma_{out} = (Z_{out} - R_{ref})/(Z_{out} + R_{ref})$,
respectively

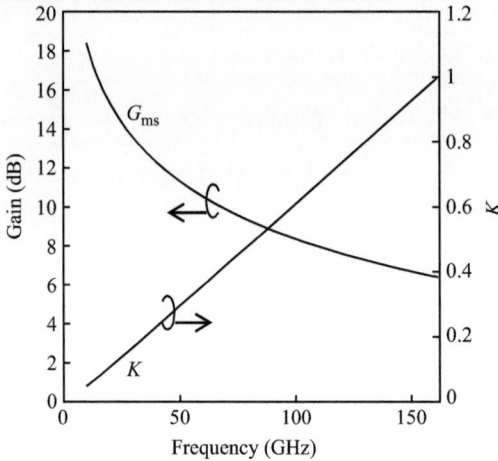

Figure 2.2   Simulated frequency dependence of the maximum stable gain (MSG)
G_{ms} and the stability factor K of a 40-nm common-source NMOSFET.
$V_{ds} = 0.9\,V$, $V_{gs} = 0.55\,V$, gate width $W = 28\,\mu m$, and $f_{max} = 260\,GHz$

When, on the other hand, $f \gtrsim f_{max}/3$, G_{ms} is typically below 10 dB (Figure 2.2), and several stages are required to build practical amplifiers (see, for example, [10,11]). In that case, one would not be as willing to trade-off gain for unconditional stability. While there are different ways of achieving high gain, such as multi-staging and feedback, we will here attempt to find the Z_S and Z_L that give high gain ($\gtrsim G_{ms}$) and actual stability, as opposed to unconditional stability. Given the fact that G_{ms} is merely a figure-of-merit, it seems important that the theoretical upper bound gain be identified. One simple application of the result would be to use it as the initial condition when performing design optimization of multistage amplifiers with a computer.

A theory of single-side matched amplifier designed for $0 < K < 1$ was developed by Edwards *et al.* [12]. They considered amplifier design in which either the source or the load side is conjugately matched. They showed that the upper bound of the single-side-matched transducer gain is $G_{msm} = 2KG_{ms}$ (maximum single-side matched gain) [12], which can exceed G_{ms} if $K > 0.5$.[1] However, design with single-side conjugate matching is restrictive in which the choice of stable terminating impedance on the mismatched side is not very wide, at times nonexistent. Also, higher transducer gain G_T can be realized if mismatch is allowed on both sides. A conditionally stable gain upper bound of $G_{mdm} = 2(K + 1)G_{ms}$ (maximum double-sided mismatched gain) is quoted in [13], but no detailed information is available about its derivation and realization.

---

[1]According to Babak [13], this result was found earlier by others and published in Russia in the Soviet Union.

In Section 2.1.2, theory of mismatched amplifier is presented, and the maximum conditionally stable gain achievable by an appropriate choice of Z_S and Z_L is derived. It turns that the upper bound is slightly different from the G_{mdm}. Section 2.1.3 shows a simple procedure for designing a high-gain ($\gtrsim G_{ms}$) amplifier with design examples. Finally, Section 2.1.4 concludes the section. Part of the results shown herein was presented in [14].

2.1.2 *Maximum conditionally stable gain*

The transducer gain G_T of a two-port depends on both Z_S and Z_L. In terms of the source and load-reflection coefficients $\Gamma_S = (Z_S - R_{ref})/(Z_S + R_{ref})$ and $\Gamma_L = (Z_L - R_{ref})/(Z_L + R_{ref})$ (Figure 2.1) [8],

$$G_T(\Gamma_S, \Gamma_L) = \frac{(1 - |\Gamma_L|^2)(1 - |\Gamma_S|^2)|S_{21}|^2}{|(1 - S_{22}\Gamma_L)(1 - S_{11}\Gamma_S) - S_{12}S_{21}\Gamma_S\Gamma_L|^2}, \tag{2.1}$$

where S_{ij} are the S parameters of the two-port (Figure 2.1). If Γ_S should be determined first as in the typical low-noise amplifier (LNA) design [6], it is a common practice to work with the available gain defined by $G_A(\Gamma_S) \equiv G_T(\Gamma_S, \Gamma_{out}^*)$ [9], where Γ_{out} is the output reflection coefficient of the two-port (Figure 2.1):

$$\Gamma_{out}(\Gamma_S) = S_{22} + \frac{S_{21}\Gamma_S S_{12}}{1 - S_{11}\Gamma_S} \tag{2.2}$$

and * denotes complex conjugate. If Γ_L should be determined first as in the typical power amplifier design [6], the operating gain $G_P(\Gamma_L) \equiv G_T(\Gamma_{in}^*, \Gamma_L)$ [9] is used instead, where Γ_{in} is the input reflection coefficient of the two-port (Figure 2.1):

$$\Gamma_{in}(\Gamma_L) = S_{11} + \frac{S_{12}\Gamma_L S_{21}}{1 - S_{22}\Gamma_L}. \tag{2.3}$$

A rough sketch of the idea for realizing high gain is as follows: a conventional stability circle on the Γ_S-plane (Smith chart) is the circle $|\Gamma_{out}| = 1$, mapped onto the Γ_S-plane using (2.2) [8,9]. The Γ_S-plane stability circle is, at the same time, a constant-G_A circle corresponding to an available gain $G_A(\Gamma_S) = G_T(\Gamma_S, \Gamma_{out}^*) \to \infty$. If, on the conjugate-matched side (the load side), the load-reflection coefficient Γ_L is appropriately "detuned" from the matched value Γ_{out}^*, $G_T(\Gamma_S, \Gamma_L)$ can assume a finite value, and the instability implied by the excessive gain ($G_A \to \infty$) can also be avoided. A parallel argument applies to the Γ_L-plane stability circle and the operating gain $G_P(\Gamma_L) = G_T(\Gamma_{in}^*, \Gamma_L)$.

As usual, graphical methods on complex reflection-coefficient planes (Γ-planes) can be used in the theoretical development. One caveat here is that Euclidean distances on a Γ-plane depend on the choice of the reference resistance R_{ref} (the resistance value at the center of a Smith chart), which is not an attribute of the two-port. Straight line segments on a Smith chart are, in general, not "immittance geodesics." If a design decision is made based on Γ-plane straight line segments as is done at times, it might not be optimal even if it appears so. This pitfall can be avoided most easily by considering immittance planes instead. With this in mind, we work here with impedances. The transducer gain in terms of the Z parameters of the two-port is [15,16]

$$G_T(Z_S, Z_L) = \frac{4R_S R_L |Z_{21}|^2}{|(Z_{11} + Z_S)(Z_{22} + Z_L) - Z_{12}Z_{21}|^2}. \tag{2.4}$$

If the two-port under consideration is unconditionally stable and therefore $K > 1$ [6–9], the values of X_S and X_L that effect simultaneous conjugate matching can be found from the condition for minimizing the denominator of (2.4), because X_S and X_L appear only in the denominator [2]. This is equivalent to the requiring that

$$X_{in} \equiv \Im(Z_{in}) = -X_S \quad \text{and} \quad X_{out} \equiv \Im(Z_{out}) = -X_L \tag{2.5}$$

as shown in Figure 2.3(a), where

$$Z_{in}(Z_L) = Z_{11} - \frac{Z_{12}Z_{21}}{Z_{22} + Z_L}, \tag{2.6}$$

$$Z_{out}(Z_S) = Z_{22} - \frac{Z_{12}Z_{21}}{Z_{11} + Z_S}. \tag{2.7}$$

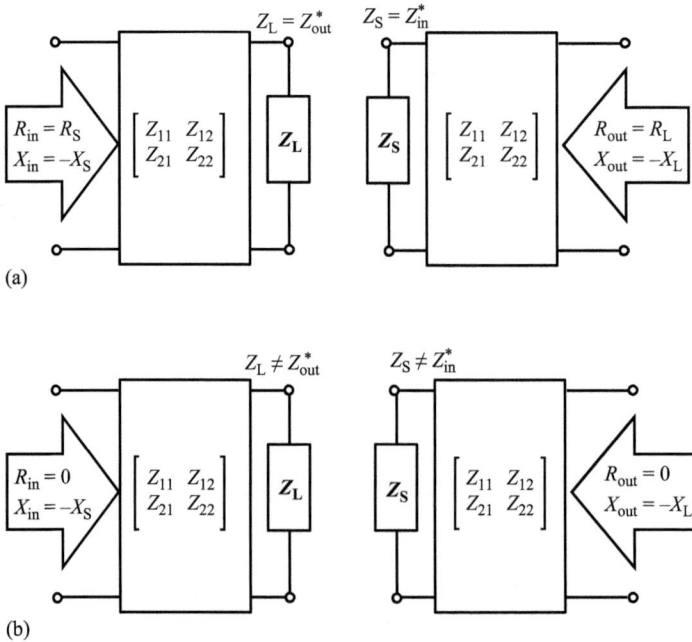

(a)

(b)

Figure 2.3   *How to choose Z_S and Z_L to achieve the theoretical upper bound transducer gain: (a) unconditionally stable case ($K > 1$) and (b) conditionally stable case ($-1 < K < 1$)*

The source and load reactance values for simultaneous conjugate matching are found to be, respectively [4,16],

$$X_S = X_{Ssc} \equiv -X_{11} + \frac{\Im(Z_{12}Z_{21})}{2R_{22}}, \tag{2.8}$$

$$X_L = X_{Lsc} \equiv -X_{22} + \frac{\Im(Z_{12}Z_{21})}{2R_{11}}, \tag{2.9}$$

where $R_{ij} = \Re(Z_{ij})$ and $X_{ij} = \Im(Z_{ij})$. The values of R_S and R_L that realize simultaneous conjugate matching can then be found by putting (2.8) and (2.9) in (2.4) and maximizing the latter with respect to R_S and R_L. The results are [16]

$$R_S = R_{Ssc} \equiv \frac{R_{11} - \Re(Z_{12}Z_{21})/(2R_{22})}{\sqrt{1 - 1/K^2}}, \tag{2.10}$$

$$R_L = R_{Lsc} \equiv \frac{R_{22} - \Re(Z_{12}Z_{21})/(2R_{11})}{\sqrt{1 - 1/K^2}}, \tag{2.11}$$

where, in terms of Z-parameters,

$$K = \frac{2R_{11}R_{22} - \Re(Z_{12}Z_{21})}{|Z_{12}Z_{21}|}. \tag{2.12}$$

The K-dependence of R_{Ssc} and R_{Lsc} is shown in Figure 2.4.

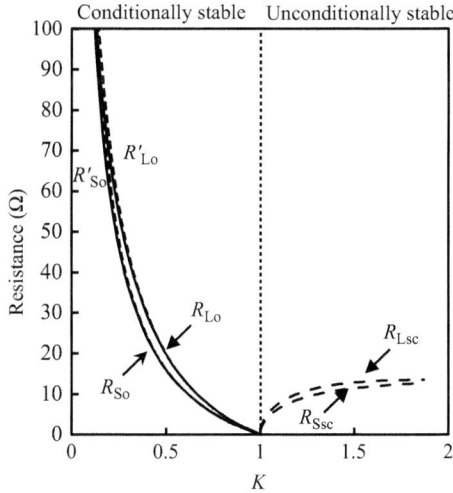

Figure 2.4 K-dependence of R_{Ssc} [Equation (2.10)], R_{Lsc} [Equation (2.11)], R_{So} [Equation (2.19)], R_{Lo} [Equation (2.20)], R'_{So} [Equation (2.33)], and R'_{Lo} [Equation (2.34)] of the 40-nm NMOSFET in Figure 2.2. When $K < 1$, $R_S > R_{So}$ and $R_L > R_{Lo}$ are required for actual stability. The plots were obtained by first calculating the frequency-dependence of the resistances and then changing the horizontal axis to K with the use of the frequency-dependence of K, shown in Figure 2.2

If $-1 < K < 1$, simultaneous conjugate matching is not possible as implied by (2.10) and (2.11). In this case, we demand for stability that [4]

$$R_{in} \equiv \Re(Z_{in}) \geq 0, \tag{2.13}$$

$$R_{out} \equiv \Re(Z_{out}) \geq 0, \tag{2.14}$$

together with (2.5), as shown in Figure 2.3(b). The equalities in (2.13) and (2.14) correspond to the border between the stable and unstable regions. The optimal (in the sense that G_T is maximized) values of X_S and X_L for the conditionally stable case turn out to be, after lengthy calculation,

$$X_S = X_{So}$$
$$\equiv X_{Ssc} + \sqrt{\frac{b^2 - 4R_{11}R_{22}R_S\left[R_{22}(R_{11} + R_S) - a\right]}{4R_{22}^2}}, \tag{2.15}$$

$$X_L = X_{Lo}$$
$$\equiv X_{Lsc} + \sqrt{\frac{b^2 - 4R_{11}R_{22}R_L\left[R_{11}(R_{22} + R_L) - a\right]}{4R_{11}^2}}, \tag{2.16}$$

where

$$a \equiv \Re(Z_{12}Z_{21}), \tag{2.17}$$

$$b \equiv \Im(Z_{12}Z_{21}). \tag{2.18}$$

Equations (2.15) and (2.16) are obviously not the same as (2.8) and (2.9), respectively, despite the use of the same equation, (2.5), in both cases. The reason for the difference is that the equality in (2.13) [Equation (2.14)] fixes the relationship between R_S and X_S [R_L and X_L]. As a result, the problem of maximizing G_T when $-1 < K < 1$ is no longer equivalent to minimizing the denominator of (2.4). The values of R_S and R_L that maximize G_T, subject to (2.13) and (2.14), are

$$R_S = R_{So} \equiv -R_{11} + \frac{\Re(Z_{12}Z_{21}) + S}{2R_{22}}, \tag{2.19}$$

$$R_L = R_{Lo} \equiv -R_{22} + \frac{\Re(Z_{12}Z_{21}) + S}{2R_{11}}, \tag{2.20}$$

where

$$S \equiv \sqrt{a^2 + 4d(b - d)}, \tag{2.21}$$

$$d \equiv T_1 R_{11} R_{22} + T_2, \tag{2.22}$$

$$T_1 \equiv -\frac{2^{1/3}(a + R_{11}R_{22})}{T}, \tag{2.23}$$

$$T_2 \equiv \frac{T}{3 \times 2^{1/3}}, \tag{2.24}$$

$$T \equiv \left[27bR_{11}^2R_{22}^2 + \sqrt{\left(27bR_{11}^2R_{22}^2\right)^2 + 4\left(3aR_{11}^2R_{22}^2\right)^3}\right]^{1/3}. \tag{2.25}$$

The K-dependence of R_{So} and R_{Lo} is plotted in Figure 2.4. The maximum conditionally stable gain thus obtained is

$$G_{mcs} = \frac{2|Z_{21}|^2 R_{11} R_{22} \left[a - 2R_{11}R_{22} + S\right]^2}{a^4 + D_1 + D_2 + D_3 + D_4}, \tag{2.26}$$

where

$$D_1 \equiv a^3 (S - 2R_{11}R_{22}), \tag{2.27}$$

$$D_2 \equiv 2b \{ b \left[T_2^2 + 2T_1 T_2 R_{11} R_{22} + \left(1 + T_1^2\right) R_{11}^2 R_{22}^2 \right]$$
$$- 2dR_{11}R_{22}S \}, \tag{2.28}$$

$$D_3 \equiv 8adR_{11}R_{22}[d + b(S - 4R_{11}R_{22})], \tag{2.29}$$

$$D_4 \equiv 2a^2 \{ - T_2^2 + 2T_1 T_2 R_{11} R_{22} + 2b(T_2 - T_1 R_{11}R_{22})$$
$$- R_{11}R_{22} \left[(T_1^2 - 1)R_{11}R_{22} + S \right] \}. \tag{2.30}$$

When $0.5 \lesssim K < 1$, the second terms in (2.15) and (2.16) are fairly small and can be neglected as shown in Figure 2.5. This leads to

$$X'_{So} \simeq X_{Ssc} \quad (0.5 \lesssim K < 1), \tag{2.31}$$

$$X'_{Lo} \simeq X_{Lsc} \quad (0.5 \lesssim K < 1). \tag{2.32}$$

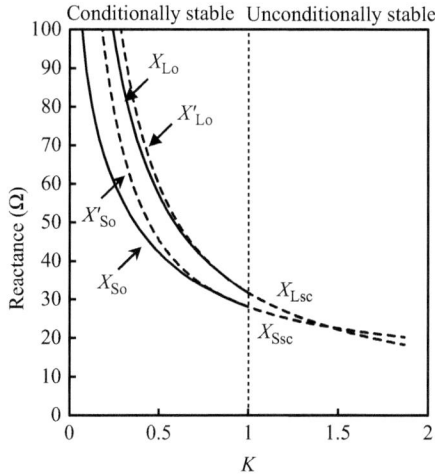

Figure 2.5   *K-dependence of X_{Ssc} [Equation (2.8)], X_{Lsc} [Equation (2.9)], X_{So} [Equation (2.15)], X_{Lo} [Equation (2.16)], X'_{So} [Equation (2.31)], and X'_{Lo} [Equation (2.32)] of the 40-nm NMOSFET in Figure 2.2*

Then, the approximate values of R_{So} and R_{Lo} are

$$R_{So} \simeq R'_{So} \quad (0.5 \lesssim K < 1)$$

$$\equiv \left[R_{11} - \frac{\Re(Z_{12}Z_{21})}{2R_{22}} \right] \left(\frac{1}{K} - 1 \right), \tag{2.33}$$

$$R_{Lo} \simeq R'_{Lo} \quad (0.5 \lesssim K < 1)$$

$$\equiv \left[R_{22} - \frac{\Re(Z_{12}Z_{21})}{2R_{11}} \right] \left(\frac{1}{K} - 1 \right). \tag{2.34}$$

R'_{So} and R'_{Lo} are also shown in Figure 2.4. The approximate maximum conditionally stable gain is given concisely by

$$G_{mcs} \simeq G'_{mcs} \equiv \frac{4R_{11}R_{22}}{|Z_{12}Z_{21}|} G_{ms} \quad (0.5 \lesssim K < 1). \tag{2.35}$$

The frequency-dependence of various gains is shown in Figure 2.6. As shown, G'_{mcs} is a good approximation to G_{mcs} for $0.5 \lesssim K < 1$. Although G_{mcs} is very close to G_{mdm} quoted in [13], there is some discrepancy as K becomes smaller. The relationship between G_{mcs} and G_{mdm} is yet to be found. But numerical sweep of Z_S and Z_L around $Z_S = R_{So} + jX_{So}$ and $Z_L = R_{Lo} + jX_{Lo}$ seemed to suggest that our analytic result, G_{mcs}, is indeed the upper bound.

The region of stability on the R_S–R_L plane for $K = 0.8$ is shown in Figure 2.7 as the unshaded area. The points on the boundary, given by (2.33) and (2.34), should never be chosen in practice, but it is important to know where the absolute edge of stability is when one cannot afford the luxury of being too safe.

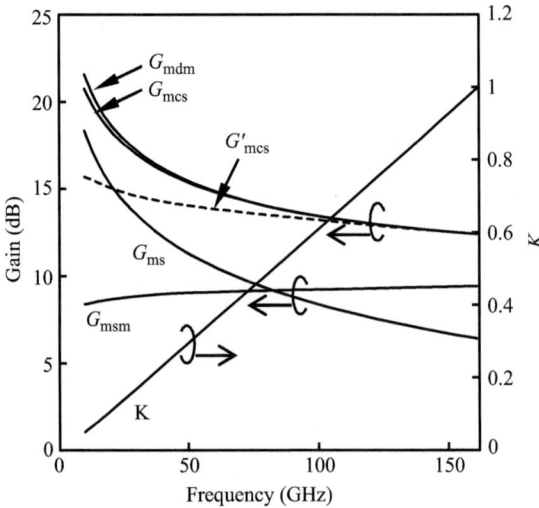

Figure 2.6   Comparison of frequency-dependence of various gains of the 40-nm NMOSFET in Figure 2.2

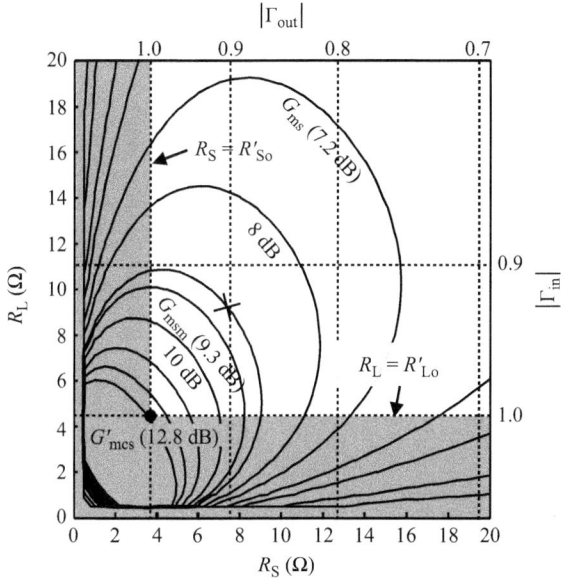

Figure 2.7 *Constant-G_T contours of the 40-nm NMOSFET in Figure 2.2 on the R_S–R_L plane at $f = 132$ GHz and $K = 0.8$. The peak gain $G'_{mcs} = 12.8$ dB is realized at the point indicated by the filled circle (•). The cross (×) indicates the point chosen for the design example shown in Figures 2.10 and 2.11*

2.1.3 Possible application to amplifier design

As a possible application of the formulas derived in the preceding subsection, we show a simple procedure for high-gain amplifier design with mismatches on both source and load sides. Since our interest is high-frequency ($f \gtrsim f_{max}/3$) design with inevitably low-gain transistors, we use the approximate expressions, (2.31) through (2.35), in the following. The examples given below are only meant to highlight the characteristics of the present approach and a typical outcome, that is, squeezing out high gain by a choice of Z_S and Z_L alone and what the result is like. No attempt is made to reach practical design solutions because that is beyond our scope. Other more elaborate procedures for designing mismatched amplifiers were reported, for example, in [17–19].

2.1.3.1 132-GHz CMOS amplifier

In this example, the target frequency is $f = 132$ GHz and the stability factor of the transistor (Figure 2.2) is $K = 0.8$. Since the optimal values of X_S and X_L are given by (2.31) and (2.32), respectively, the main task is to choose the values of R_S and R_L appropriately considering the trade-off between gain and stability margin. The constant-G_T contours on the R_S–R_L plane (Figure 2.7) facilitate the choice. The MSG at this frequency is $G_{ms} = 7.2$ dB and we set the target as $G_T = 9$ dB. The values of

R_S and R_L should be far enough from the stability borders R'_{So} and R'_{Lo}. We chose $R_S = 7.7\,\Omega$ and $R_L = 9.8\,\Omega$, marked with a cross in Figure 2.7.

To show the connection of the above with the familiar design procedure that makes use of Smith charts [6–9], Γ_S- and Γ_L-plane Smith charts, together with stability and constant-G_T circles, are presented in Figures 2.8 and 2.9. A Z_S-first high-gain amplifier design with Smith charts could proceed as follows:

1. Determine X_S and X_L using (2.31) and (2.32). These are shown in Figures 2.8 and 2.9 as dashed lines.
2. Choose a Γ_S on the dashed line on the Γ_S-plane (Figure 2.8) such that it is far enough from the stability circle and that the gain is sufficient, thereby fixing Z_S and $Z_{out}(Z_S)$ [Equation (2.7)].
3. Likewise, choose a Γ_L on the dashed line on the Γ_L-plane (Figure 2.9), thereby fixing Z_L and $Z_{in}(Z_L)$ [Equation (2.6)].

The schematic diagram of the designed amplifier with simple L-type impedance-transforming networks is shown in Figure 2.10(a). S parameters of the amplifier are shown in Figure 2.10(b). An $|S_{21}|$ greater than the G_{ms} is successfully achieved, but the input and output matching is rather poor at $f = 132\,\text{GHz}$. This is because the chosen values of R_S and R_L tend to be significantly lower than $50\,\Omega$ when $0.5 < K < 1$, as implied by Figure 4.[2] Also, out-of-band $|S_{11}|$ slightly exceeds unity in Figure 2.10(b). The reason is essentially the same, and the L-type impedance-transforming network is not good enough. As a quick remedy, the balanced amplifier configuration [6,7,20] is adopted in Figure 2.11(a). Figure 2.11(b) shows significantly better matching

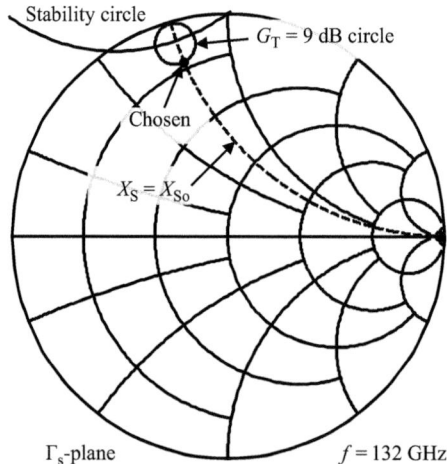

Figure 2.8 Γ_S-plane Smith chart of the 40-nm NMOSFET in Figure 2.2 at
f = 132 GHz. Stable region is outside the stability circle

---

[2]Note that $R_S > R'_{So}$ and $R_L > R'_{Lo}$ in the design.

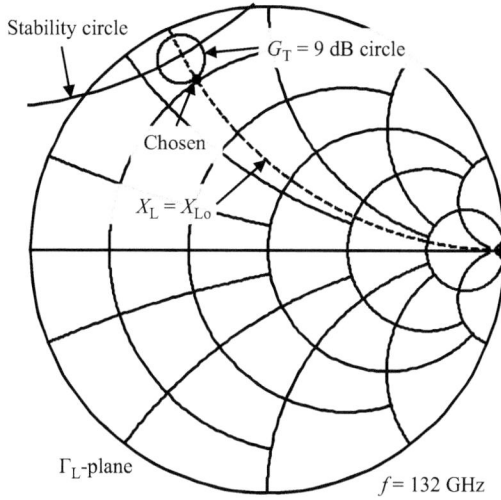

Figure 2.9 Γ_L-plane Smith chart of the 40-nm NMOSFET in Figure 2.2 at
$f = 132$ GHz. Stable region is outside the stability circle

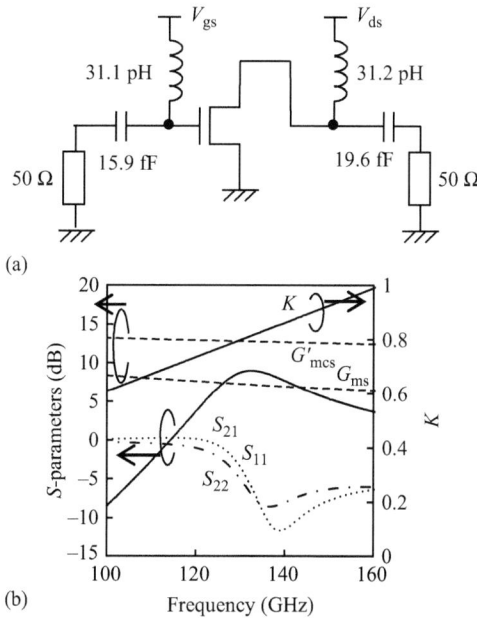

Figure 2.10 (a) 132 GHz common-source amplifier. $R_S = 7.7\,\Omega$ and $R_L = 9.8\,\Omega$.
(b) Simulated S parameters of the designed amplifier

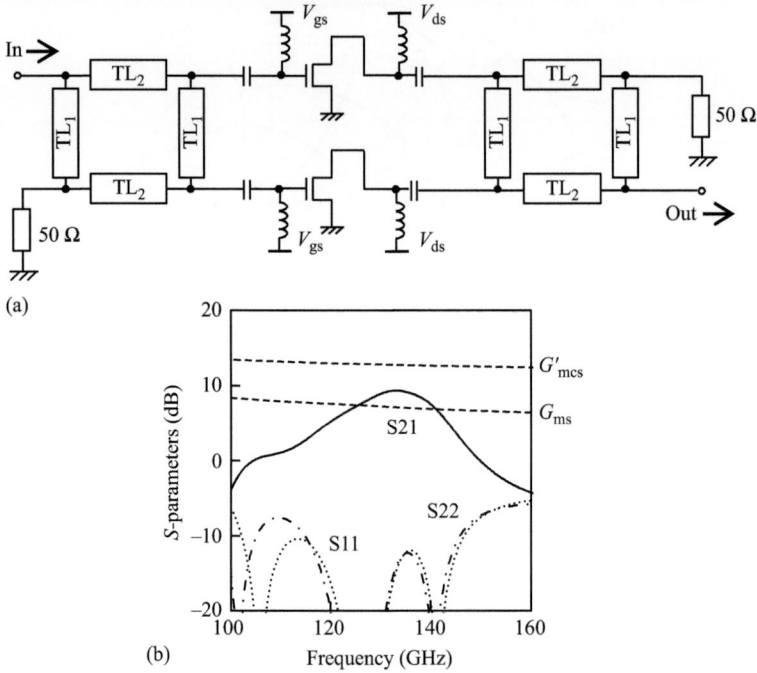

Figure 2.11   *(a) 132 GHz balanced amplifier built with the common-source amplifiers shown in Figure 2.10 and branch-line couplers. TL_1 and TL_2 are $\lambda/4$ transmission lines (TLs) of characteristic impedance 50 and 35.4 Ω. (b) Simulated S parameters of the designed amplifier*

properties. Of course, the cost of introducing 3-dB hybrids can be significant. In any case, the important point here is that great care is required to improve input and output matching.

2.1.3.2   60-GHz amplifier design with parameter dispersion

In this example, we use measured S parameters of 12 NMOSFETs on 12 nominally the same 40-nm complementary metal-oxide-semiconductor (CMOS) chips. This is to demonstrate how device parameter dispersion might appear during the design. The design procedure itself is the same as the previous example. Here the MOSFETs' f_{max} is deliberately lowered by applying somewhat low bias voltages of $V_{ds} = 0.4$ V and $V_{gs} = 0.725$ V. Biasing MOSFETs like this is highly beneficial in terms of power efficiency and can have practical utility [21]. The frequency dependence of measured G_{ms} is shown in Figure 2.12. Γ_S-plane stability circles and constant-gain ($G_T = 8$ dB) circles are shown in Figure 2.13 as an example. The results include measurement uncertainties and spread of intrinsic device characteristics. The schematic diagram and the gain of the designed amplifier are shown in Figures 2.14 and 2.15, respectively. The message here is that when parameter dispersion is taken into consideration, the target G_T has to be lowered.

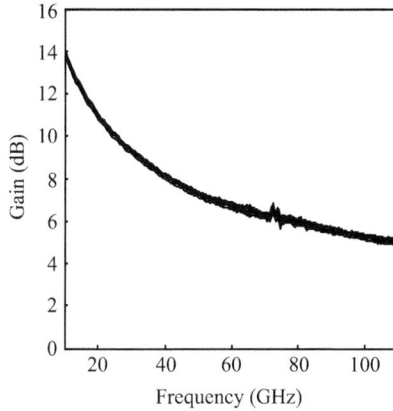

Figure 2.12 *Measured frequency-dependence of G_{ms} of 12 NMOSFETs. $V_{ds} = 0.4\,V$, $V_{gs} = 0.725\,V$, gate length $L = 40\,nm$, and gate width $W = 40\,\mu m$*

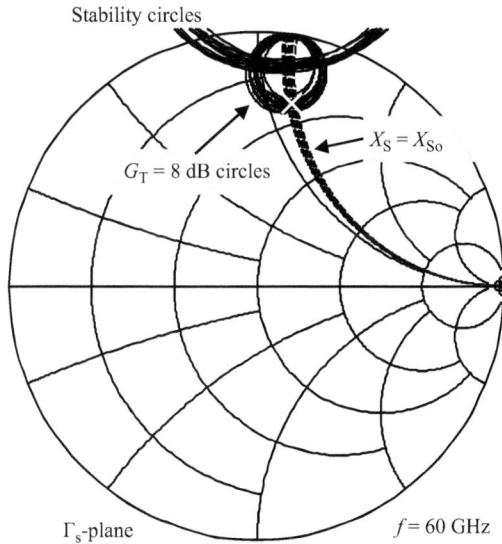

Figure 2.13 *Stability circles of 12 NMOSFETs (solid lines) on the Γ_S-plane at $f = 60\,GHz$. Constant-reactance circles corresponding to the 12 values of X_{So} are also shown (dashed lines). The white cross indicates the point chosen for the design*

2.1.4 Summary and discussion

We analytically derived the maximum conditionally stable gain G_{mcs} [Equation (2.26)]. It is the theoretical upper bound of the stable transducer gain and is greater than the so-called MSG given by $G_{ms} = |S_{21}/S_{12}|$ (Figure 2.6). G_{mcs} is realized by setting the

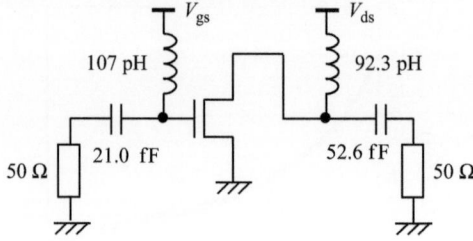

Figure 2.14   Schematic diagram of the designed 60-GHz amplifier

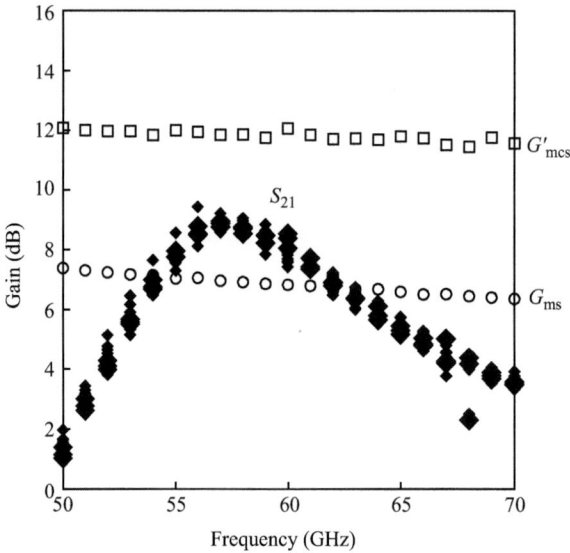

Figure 2.15   Frequency dependence of S_{21} (filled diamonds) of 12 amplifiers (Figure 2.14). G_{ms} (○) and G'_{mcs} (□) of a typical device are also plotted for comparison

source and load impedance values to $Z_S = R_{So} + jX_{So}$ [Equations (2.19) and (2.15)] and $Z_L = R_{Lo} + jX_{Lo}$ [Equations (2.20) and (2.16)], respectively. We also presented simpler approximate expressions valid for $0.5 \lesssim K < 1$.

As a possible application of the above, we showed design examples of conditionally stable small-signal high-gain ($G_T \gtrsim G_{ms}$) amplifiers. The motivation for using such a technique was that the number of gain stages required to achieve a target overall gain could be reduced. However, amplifiers designed this way tend to have poor input and output matching, as suggested by Figure 2.4, due to the low values of $R_S = \Re(Z_S)$ and $R_L = \Re(Z_L)$ required. Therefore whether the design technique presented in Section 2.1.3 can be a practical solution to high-gain amplifier design might be somewhat questionable. A much safer application of the derived results will be to use the upper bound condition as the "high-gain" initial condition for computer-aided design optimization.

The above difficulty, incidentally, gives us the solid rationale for going for the near-f_{max} gain-boosting technique of much greater complexity: feedback.[3] With feedback, it is, in principle, possible to make the amplifier unconditionally stable ($K > 1$) and boost the gain to the theoretical upper bound value for the K, derived in [22].

2.2   Gain and noise optimization of small-signal amplifier [23]

At terahertz frequencies, many ($\gtrsim 5$) gain stages are required to build amplifiers due to the limited MOSFET gain ($\lesssim 5$ dB). However, in general, optimization of many-stage amplifiers is difficult because the design of each matching network cannot be done independently of others due to non-negligible reverse isolation (S_{12}) of MOSFETs. Accordingly, computational optimization is necessary.

In our approach, small-signal transfer functions of all constituent devices with geometrical parameters are first modeled. Then, the overall transfer function of the amplifier is calculated. After giving initial parameter values (e.g., by [14]) to the constituent devices, a conjugate gradient engine maximizes an objective function as shown in Figure 2.16 [24]. For ultrahigh-speed communication at 10 Gbit/s or faster, amplifiers with flat gain and flat group delay over a very wide frequency range are required. In our optimization procedure, interstage matching networks are optimized first, and then the input and output matching networks are optimized, as shown in Figure 2.17.

Note that the objective function can be defined to search for lower noise figure (NF) or for higher gain. The optimization procedure can also be tuned depending on the objective. The gain can be maximized by realizing simultaneous conjugate matching at the input and the output if the amplifier is unconditionally stable. On the other hand, NF (or equivalently, noise factor) is minimized when the source impedance has a certain optimal value. Generally, conditions of conjugate power matching and minimum NF are not the same. Therefore, NF of the first gain stage is usually minimized in microwave LNAs since NFs of the succeeding stages do not affect the overall noise performance very much. However, the situation is somewhat different in terahertz amplifiers. The typical gain per stage at terahertz frequencies is so low that the NF of the second and even the third stage may affect the overall noise performance of a multistage amplifier. Generally, the higher the gain of the first gain stage, the less the impact of the NFs of the later stages on the overall NF [25]. Therefore, conjugate power matching at the input port of the first stage might give a smaller NF than does the ordinary "noise matching" strategy.

We used the four-stage common-source amplifier, shown in Figure 2.18, as a test vehicle and tested two design strategies in computational optimization. The design targets were a center frequency of 135 GHz and a fractional bandwidth of over 20%. The

---

[3]Unnecessary complexity in design should, of course, be avoided as much as possible.

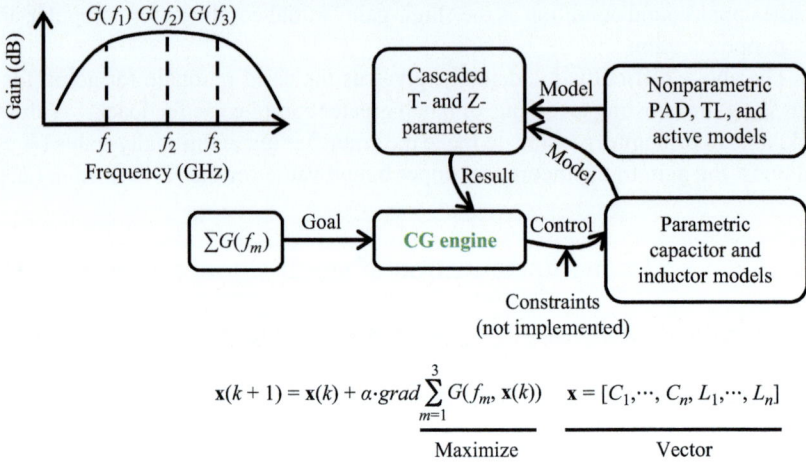

$$\mathbf{x}(k+1) = \mathbf{x}(k) + \alpha \cdot grad \underbrace{\sum_{m=1}^{3} G(f_m, \mathbf{x}(k))}_{\text{Maximize}} \underbrace{\mathbf{x} = [C_1, \cdots, C_n, L_1, \cdots, L_n]}_{\text{Vector}}$$

Figure 2.16   *Optimization process. The conjugate-gradient (CG) engine searches for a flat and wideband characteristic with an achievable gain for the amplifier. Copyright 2015 IEICE. Reproduced, with permission, from [11]*

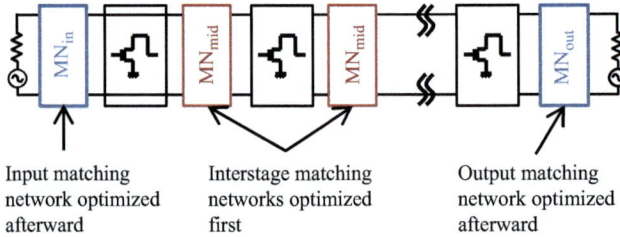

Figure 2.17   *Interstage matching networks are optimized first. Then, input and output matching networks are optimized. Copyright 2015 IEICE. Reproduced, with permission, from [11]*

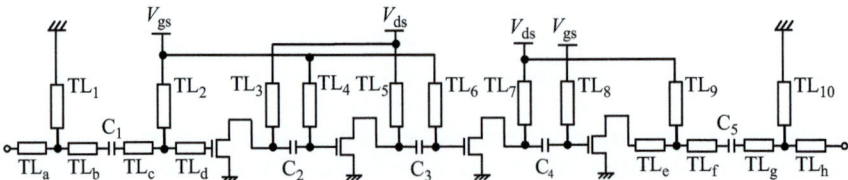

Figure 2.18   *Four-stage 135-GHz amplifier used in performance optimization test. Copyright 2015 IEICE. Reproduced, with permission, from [11]*

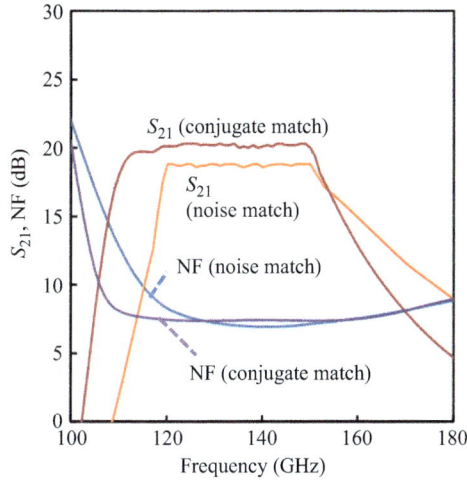

Figure 2.19 *S_{21} and noise figures of the amplifier in Figure 2.18 when the conjugate-matching and the noise-matching strategies are adopted in optimization. Copyright 2015 IEICE. Reproduced, with permission, from [11]*

results of applying conjugate matching and noise matching are shown in Figure 2.19. The gain (S_{21}) of the noise-matched case is lower than the conjugate-matched case as expected. But the NF of the noise-matched case is not significantly better than the conjugate-matched case. This is in stark contrast with ordinary low-GHz amplifiers.

As illustrated in this example, the conventional low-noise design might not be optimal for terahertz design.

2.3 Gain-boosting by feedback

2.3.1 Introduction

Empirically, operation frequency f of production level, reliable microwave circuits satisfies $f \lesssim f_{max}/4$ or $f \lesssim f_{max}/5$, where f_{max} is the transistor's maximum operation frequency [5]. However, research into amplifiers operating at $f > f_{max}/2$ is becoming active [26,27]. At such high frequencies, the MAG [9] G'_{ma} of the transistor is rather low. Since G'_{ma} is the upper bound of the gain of a plain single-stage amplifier, gain boosting with feedback is desirable. While the MAG G_{ma} of a composite amplifier with feedback (Figure 2.20) can, in theory, be brought to infinity at $f < f_{max}$ with a certain lossy or nonreciprocal embedding network [28], such an extreme configuration is of little practical interest for design of amplifiers (as opposed to oscillators) for stability reasons. In what follows, we assume the embedding network to be lossless and reciprocal, for which the highest G_{ma} achievable is bounded. Under this assumption, Mason's unilateral gain U [29] of the composite amplifier is equal to that of the core

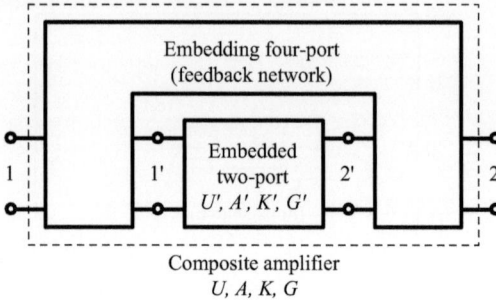

Figure 2.20   *Composite two-port amplifier consisting of an embedded core two-port amplifier and an embedding four-port network. Variables with primes refer to the properties of the core two-port, and those without primes refer to the properties of the composite two-port. U is Mason's unilateral gain (2.36). A is the transfer-parameter ratio (2.38). K is the stability factor (2.37). G is the gain (2.40). Copyright 2014 IEICE. Reproduced, with permission, from [22]*

amplifier ($U = U'$), regardless of the details of the embedding network. As will be shown shortly, this greatly facilitates theoretical analysis. Although actually realizable embedding networks at near-f_{max} frequencies would inevitably be lossy, the insights gained from the study of lossless reciprocal embedding is still useful in knowing the performance upper bound and setting practical design targets.

Here, variables with primes denote properties of the core two-port. Variables without primes denote properties of the composite two-port (Figure 2.20).

2.3.2   Gain and stability of feedback amplifier [22]

Mason's U can be written in terms of the stability factor K [9] and the transfer-parameter ratio A, defined by (2.38), as [30,31]

$$U = \frac{|A - 1|^2}{2K|A| - 2\Re(A)},\tag{2.36}$$

$$K = \frac{1 - |S_{11}|^2 - |S_{22}|^2 + |\det \mathsf{S}|^2}{2|S_{12}S_{21}|},\tag{2.37}$$

$$A \triangleq \frac{S_{21}}{S_{12}} = \frac{Y_{21}}{Y_{12}} = \frac{Z_{21}}{Z_{12}} = \det \mathsf{T}.\tag{2.38}$$

Here S_{ij}, Y_{ij}, and Z_{ij} are, respectively, S, Y, and Z parameters of the composite two-port, and S and T are its S and T matrices. Note that A is complex. It can be shown from (2.36) that [31]

$$U = \frac{G|A - 1|^2}{|A|^2 - 2G \cdot \Re(A) + G^2},\tag{2.39}$$

where G is a gain defined by

$$G \triangleq \frac{|A|}{K + \sqrt{K^2 - 1}}.$$ (2.40)

Obviously,

$$|G| = \begin{cases} G = G_{\text{ma}} & (K \geq 1) \\ |A| = G_{\text{ms}} & (-1 < K \leq 1) \end{cases},$$ (2.41)

where G_{ms} is the MSG [9].

Lossless reciprocal embedding includes such practical operations as reactive parallel and series embedding (by Y and Z matrices, respectively), lossless cascading (by T matrix), and port permutation (by ideal transformer). As mentioned earlier, Mason's unilateral gain U' of the core two-port is invariant to lossless reciprocal embedding [29], that is, $U = U'$. But A' (and K') are transformed into A (and K) by the same embedding unless it is lossless cascading (Figure 2.21). For chosen values of U and A, the composite two-port's G can be found from (2.39). By appropriately designing an embedding four-port, A' is transformed into a right value of A, and a desired G (and K) are obtained. U can be lowered if necessary without affecting A by lossy cascading (Figure 2.21).

Singhakowinta and Boothroyd theoretically studied the performance of feedback amplifiers with lossless reciprocal embedding and derived from (2.39) an important relationship among U, A, and G_{ma} of the composite amplifier [31,32]:

$$\sqrt{\frac{G_{\text{ma}}}{U}} = \left| \frac{A - G_{\text{ma}}}{A - 1} \right| \quad (K \geq 1).$$ (2.42)

In [31], the authors assumed $|A| \gg 1$, which greatly simplified the subsequent analysis. But since our interest is near-f_{max} design, we do not make that approximation.

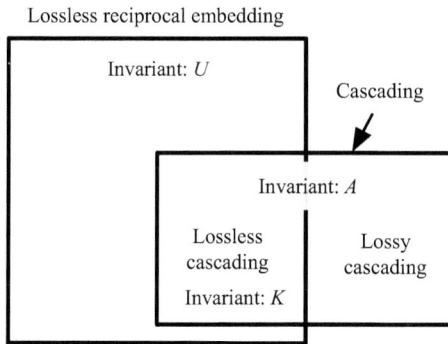

Lossless reciprocal embedding

Invariant: U

Cascading

Invariant: A

Lossless cascading

Lossy cascading

Invariant: K

Figure 2.21 *Two-port transformation operations (by embedding as in Figure 2.20) and associated invariants. Lossless reciprocal embedding leaves U' invariant, leading to $U = U'$. Cascading leaves A' invariant, leading to $A = A'$. Lossless cascading leads to $K = K'$, $U = U'$, and $A = A'$. Copyright 2014 IEICE. Reproduced, with permission, from [22]*

Exact analytic treatment valid for all values of A (provided $K \geq 1$) is actually possible. For given values of U and G_{ma}, (2.42) represents a constant-G_{ma} circle (of Apollonius) on the complex A-plane. The equation of the circle is given by

$$\left[A_{\text{R}} + \frac{G(U-1)}{G-U} \right]^2 + A_{\text{I}}^2 = \left(\frac{G-1}{G-U} \right)^2 UG, \tag{2.43}$$

where $A_{\text{R}} = \Re(A)$ and $A_{\text{I}} = \Im(A)$. Alternatively, we can introduce $\lambda \equiv 1/A$ and work in the inverse gain space [31], where the unilateral case ($S_{12} = 0$) corresponds to the origin, $\lambda = 0$, on the λ-plane. A constant-G_{ma} circle on the A-plane is then mapped onto a circle on the λ-plane. The equation of the constant-G_{ma} circle for a given U is found to be

$$\left(\lambda_{\text{R}} - \frac{U-1}{UG_{\text{ma}}-1} \right)^2 + \lambda_{\text{I}}^2 = \frac{(G_{\text{ma}}-1)^2}{(UG_{\text{ma}}-1)^2} \cdot \frac{U}{G_{\text{ma}}}, \tag{2.44}$$

where $\lambda_{\text{R}} = \Re(\lambda) = \Re(A)/|A|^2$ and $\lambda_{\text{I}} = \Im(\lambda) = -\Im(A)/|A|^2$.

Not every point on the λ-plane is of interest because some might not satisfy $K \geq 1$. To see if a point on the λ-plane is unconditionally stable, K should be expressed in terms of U and A (or λ). From (2.36),

$$K = \frac{|A|^2 + 2(U-1)\Re(A) + 1}{2|A|U} \tag{2.45}$$

$$= \frac{\lambda_{\text{R}}^2 + \lambda_{\text{I}}^2 + 2(U-1)\lambda_{\text{R}} + 1}{2U\sqrt{\lambda_{\text{R}}^2 + \lambda_{\text{I}}^2}}. \tag{2.46}$$

(2.45) and (2.46) suggest that K of the composite two-port is controllable through the choice of $A = 1/\lambda$ and U. For example, an unstable core amplifier with $K' < 1$ could be stabilized ($K > 1$), or a stable core with $K' > 1$ could produce a higher gain ($G_{\text{ma}} > G'_{\text{ma}}$) by trading-off some stability ($1 < K < K'$).

Some constant-G_{ma} circles are shown in Figures 2.22 and 2.23 together with constant-K contours. Constant-G_{ma} circles intersect the boundary of unconditional stability ($K = 1$) tangentially, and the entire circles are in the unconditionally stable region. This is reasonable, for G_{ma} should exist for any value of $K \geq 1$. Although [31] suggested that constant-G_{ma} circles (actually, arcs) intersect the $K = 1$ contour nontangentially, the nontangential intersection appears to be an artifact of their approximation, $|A| \gg 1$. This difference leads to some new findings. The $K = 1$ contour gives a teardrop shape (Figure 2.22) as opposed to a parabola in the approximate treatment [31].

The highest value of G_{ma} is achieved at the leftmost point on the $K = 1$ contour (Figures 2.22 and 2.23), which is on the real axis. From (2.39) with $A = -\Re(A)$ and $G = |A|$, we get

$$G^2 - 2(2U-1)G + 1 = 0, \tag{2.47}$$

which leads to the maximum gain reported in [32]

$$G_{\text{max}} = 2U - 1 + 2\sqrt{U(U-1)}. \tag{2.48}$$

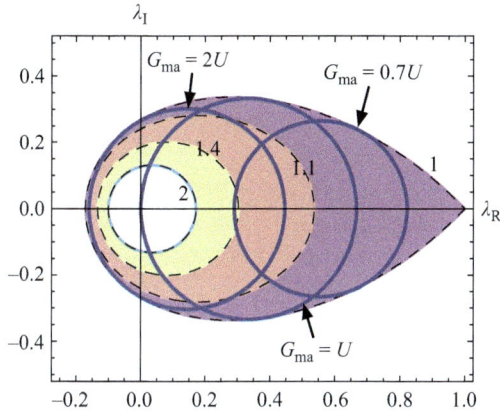

Figure 2.22 Solid circles: constant-G_{ma} contours on the λ-plane with $G_{ma} = 2U$, U, 0.7U for U = 2 (3 dB), which is a typical near-f_{max} value. Dashed closed curves: constant-K contours for K = 1, 1.1, 1.4, 2. Inside the teardrop shape corresponding to K = 1 is the unconditionally stable region, in which G_{ma} can be defined. Copyright 2014 IEICE. Reproduced, with permission, from [22]

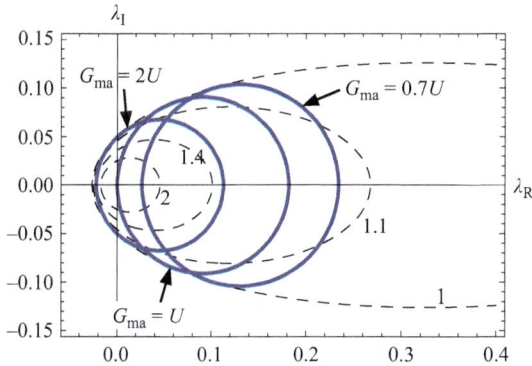

Figure 2.23 Solid circles: constant-G_{ma} contours on the λ-plane with $G_{ma} = 2U$, U, 0.7U for U = 10 (10 dB). Dashed (closed) curves: constant-K contours for K = 1, 1.1, 1.4, 2. Copyright 2014 IEICE. Reproduced, with permission, from [22]

The leftmost point on the $K = 1$ contour (on the real axis) gives $G_{\max}$ (Figures 2.22 and 2.23).

Given a transistor with $A' = S'_{21}/S'_{12}$, design of an unconditionally stable feedback amplifier with lossless reciprocal embedding is a problem of choosing a value of $\lambda = 1/A$ that gives a target value of G_{ma} and somehow synthesizing an embedding four-port that transforms λ' into the chosen λ. For a given target value of G_{ma}, the resulting composite amplifier should be as stable as possible. Since points on

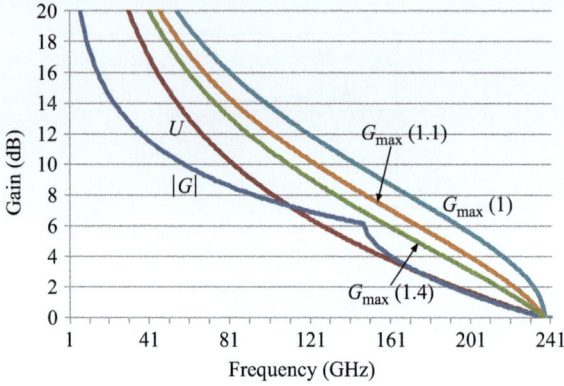

Figure 2.24   *Frequency dependence of U, |G|, and $G_{max}(K)$ of a 40-nm NMOSFET with $f_{max} \simeq 237$ GHz (simulation). Copyright 2014 IEICE. Reproduced, with permission, from [22]*

a constant-G_{ma} circle assume the highest K-value at the left point of intersection with the real axis (Figures 2.22 and 2.23), that point is the optimal point. To make $G_{ma} > U$, λ (and A) should lie on the negative part of the real axis. In that case, (2.39) with $\Re(A) = -|A|$ gives the highest gain achievable for a desired value of $K(\geq 1)$:

$$G_{max}(K) = \frac{(K+1)U - 1 + \sqrt{[(K+1)U - 1]^2 - 1}}{K + \sqrt{K^2 - 1}}, \tag{2.49}$$

which reduces to (2.48) as $K \to 1$. As shown in Figure 2.24, even a slight improvement in stability (greater value of K) results in considerable reduction in $G_{max}(K)$. It approaches U at $\lambda = 0$ (the unilateral condition) as $K \to \infty$.

A' (and λ') can be transformed into A (and λ) by applying, for example, simple embedding networks shown Figure 2.25 a few times [31]. The feedback configurations shown in Figure 2.25(a) and (b) give, respectively,

$$A = \frac{Y'_{21} - jB_F}{Y'_{12} - jB_F} = \frac{A' - jB_F/Y'_{12}}{1 - jB_F/Y'_{12}} \quad \text{[Figure 2.25(a)]}, \tag{2.50}$$

$$A = \frac{Z'_{21} + jX_F}{Z'_{12} + jX_F} = \frac{A' + jX_F/Z'_{12}}{1 + jX_F/Z'_{12}} \quad \text{[Figure 2.25(b)]}. \tag{2.51}$$

In [27], a near-f_{max} amplifier employing transmission line feedback was presented. A section of lossless transmission line with a length ℓ can be represented by the Π network shown in Figure 2.26 with $Y_1 = jY_0 \tan(\beta\ell/2)$ and $Y_2 = -jY_0 \csc\beta\ell$, where Y_0 and β are the line's characteristic conductance and phase constant, respectively. Since the diagonal elements of $\mathbf{Y_F}$ do not affect A, transmission line feedback can be understood in a similar manner to Figure 2.25(a). Amplifiers employing lossless transformer feedback can also be analyzed and designed within the framework of the present theory.

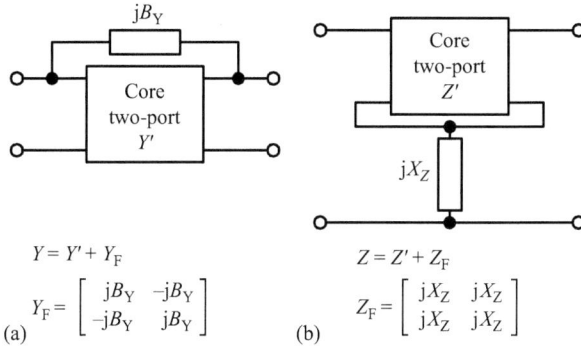

$$Y = Y' + Y_F$$

$$Y_F = \begin{bmatrix} jB_Y & -jB_Y \\ -jB_Y & jB_Y \end{bmatrix}$$

(a)

$$Z = Z' + Z_F$$

$$Z_F = \begin{bmatrix} jX_Z & jX_Z \\ jX_Z & jX_Z \end{bmatrix}$$

(b)

Figure 2.25   *Simple embedding networks that transform $\lambda' = S'_{12}/S'_{21}$ to $\lambda = S_{12}/S_{21}$: (a) shunt–shunt or Y feedback and (b) series–series or Z feedback. Copyright 2014 IEICE. Reproduced, with permission, from [22]*

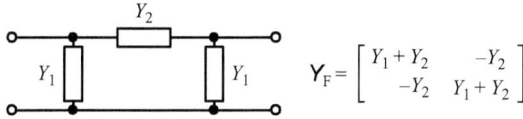

$$Y_F = \begin{bmatrix} Y_1 + Y_2 & -Y_2 \\ -Y_2 & Y_1 + Y_2 \end{bmatrix}$$

Figure 2.26   *Π network as an embedding network. $\mathbf{Y} = \mathbf{Y'} + \mathbf{Y_F}$. Copyright 2014 IEICE. Reproduced, with permission, from [22]*

2.3.3   Gain boosting by lossless reciprocal feedback [22]

Transistors are usually regarded as active devices. However, they become passive beyond the unity-power-gain frequency f_{max} [33], and all amplifiers inevitably operate below that frequency. Design of amplifiers that operate roughly above $f_{max}/2$ (near-f_{max} amplifiers) is possible but is quite a challenge due to the limited gain available at such frequencies. The recent interest in terahertz technology has led to reports on near-f_{max} gain-boosted feedback amplifiers with impressive performance [26,27,34]. The designs in these papers are based on an optimal condition expressed in terms of Y parameters [26,35]. However, this mathematical condition says very little about how to actually synthesize a feedback network. References [26,27,34] adopted a simple inductive shunt–shunt feedback configuration [16], without discussing whether it indeed realized the optimal condition. As will be shown shortly, this configuration does not always allow the MAG, G_{ma}, of the feedback amplifier to be brought close enough to the theoretical upper bound value given by (2.49) with $K = 1$.

    To answer the question of how to synthesize a feedback network for better performance, we present a graphical approach to designing gain-boosted near-f_{max} amplifiers. Such a graphical method was originally proposed by Singhakowinta and Boothroyd [31]. However, they made some simplifying assumptions in their formulation, which become invalid near the f_{max}. Later, rigorous theory applicable also at higher frequencies was established [22] as was explained in Section 2.3.2. Our graphical method is based on this theory.

The essential idea for gain boosting in [26,27,34] is to embed a core two-port amplifier (a transistor) in an appropriate lossless reciprocal embedding (feedback) network (Figure 2.27). While losses in actual components constituting the embedding network must eventually be taken into consideration, this idealized treatment gives a reasonable approximate design solution. Lossless reciprocal embedding is special in which Mason's unilateral gain of the core (U') and that of the composite two-port (core + embedding network) (U) are equal ($U' = U$) [29]. In contrast, the MAG (G'_{ma} and G_{ma}) and the stability factor [9] (K' and K) of the core and the composite two-port, respectively, may be different ($G'_{ma} \neq G_{ma}$, $K' \neq K$), depending on the details of the embedding network. The values of G_{ma} and K can be controlled (interdependently) by appropriately designing the embedding network [22]. In this sense, the embedding network transforms G'_{ma} (K') to G_{ma} (K). The highest possible value of G_{ma} achievable for a given stability $K(\geq 1)$ is given by (2.49). The upper bound value, $G_{max}(1)$, was reported earlier in [32]. A good near-f_{max} design goal will be to synthesize a feedback network that makes G_{ma} as close to $G_{max}(K)$ as possible with K being close to, but slightly greater than, 1.

Design of feedback network can be done graphically by considering the complex inverse gain plane (λ-plane), spanned by $\lambda_R = \Re(\lambda)$ and $\lambda_I = \Im(\lambda)$ [22,31], where

$$\lambda^{-1} \triangleq \frac{S_{21}}{S_{12}} = \frac{Y_{21}}{Y_{12}} = \frac{Z_{21}}{Z_{12}}. \tag{2.52}$$

Here S_{ij}, Y_{ij}, and Z_{ij} are, respectively, S, Y, and Z parameters of the two-port in question (core or composite). $|\lambda^{-1}|$ equals the MSG, G_{ms} [9]. When the two-port is

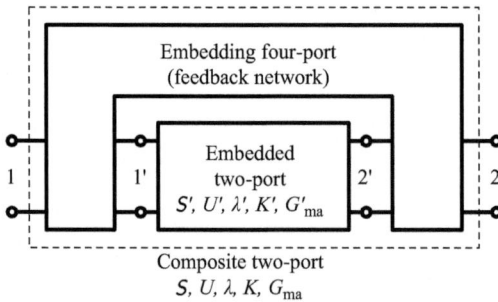

Figure 2.27   *Feedback amplifier composed of a core two-port (transistor) and an embedding four-port. Variables with a prime denote properties of the core two-port, and those without denote the properties of the composite two-port. **S** is two-port's S matrix, U is Mason's unilateral gain, λ is the complex inverse gain, (2.52), K is the stability factor, (2.53) and G_{ma} is the MAG, (2.54)*

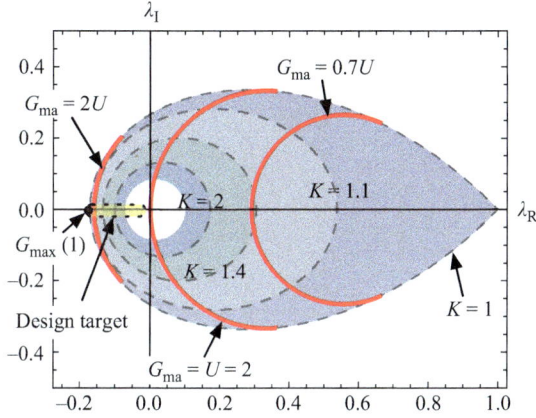

Figure 2.28   *λ-plane MAG-K chart for U = 2 (3 dB). Dashed closed curves are constant-K contours for K = 1, 1.1, 1.4, 2. Red arcs are constant-G_{ma} contours for G_{ma} = 2U, U, 0.7U. The yellow strip is a high-gain design target. The theoretical maximum gain, $G_{max}(1)$, is realized at a single point. Copyright 2016 IEEE. Reproduced, with permission, from [37]*

unconditionally stable ($K \geq 1$ and $|\det \mathbf{S}| \leq 1$ [9]), a point on a λ-plane has associated values of K and G_{ma}, for these are functions of λ and U as follows [22]:

$$K(\lambda, U) = \frac{\lambda_R^2 + \lambda_I^2 + 2(U-1)\lambda_R + 1}{2U\sqrt{\lambda_R^2 + \lambda_I^2}}, \tag{2.53}$$

$$G_{ma}(\lambda, U) = \frac{|\lambda^{-1}|}{K + \sqrt{K^2 - 1}} \quad (K \geq 1). \tag{2.54}$$

Once U is given, (2.53) and (2.54) are, respectively, the equations of constant-K and constant-G_{ma} contours (arcs in the case of G_{ma}) on a λ-plane. Figure 2.28 shows a λ-plane *MAG-K chart* for $U = 2$, which is a typical near-f_{max} value. Only the colored region (and within) is unconditionally stable ($K \geq 1$) and is of interest. Note that the shapes of constant-K contours depend on the value of U [22]. We can see from the contours in Figure 2.28 that the yellow region (strip) gives high gain ($G_{ma} > U$) and best stability (largest K) for a given gain. A caveat is that G_{ma} is sensitive to circuit parameter variations in the high-gain region because constant-G_{ma} contours are dense there [36].

2.3.4   Graphical design of feedback network [37]

Let λ′ denote the core two-port's value of the complex inverse gain, (2.52), and let λ denote that of the composite two-port. When some feedback is applied to the core, λ′ is transformed to λ. Simplest examples of feedback networks are shown in Figure 2.25. The transformation formulas for Figure 2.25(a) and (b) are, respectively,

$$\lambda = \frac{\lambda' - jB_Y/Y'_{21}}{1 - jB_Y/Y'_{21}} \quad \text{[Y feedback, Figure 2.25(a)],} \tag{2.55}$$

$$\lambda = \frac{\lambda' + jX_Z/Z'_{21}}{1 + jX_Z/Z'_{21}} \quad \text{[Z feedback, Figure 2.25(b)].} \tag{2.56}$$

When B_Y and X_Z are swept, the loci resulting from (2.55) and (2.56) are circles on the λ-plane. The equations of the circles are, respectively [37],

$$\left[\lambda_R - \frac{r\lambda'_R - q\lambda'_I + r}{2r}\right]^2 + \left[\lambda_I - \frac{q\lambda'_R + r\lambda'_I - q}{2r}\right]^2$$

$$= \left[\frac{r\lambda'_R - q\lambda'_I + r}{2r}\right]^2 + \left[\frac{q\lambda'_R + r\lambda'_I - q}{2r}\right]^2 - \frac{r\lambda'_R - q\lambda'_I}{r}, \tag{2.57}$$

$$\left[\lambda_R - \frac{c\lambda'_R - d\lambda'_I + c}{2c}\right]^2 + \left[\lambda_I - \frac{d\lambda'_R + c\lambda'_I - d}{2c}\right]^2$$

$$= \left[\frac{c\lambda'_R - d\lambda'_I + c}{2c}\right]^2 + \left[\frac{d\lambda'_R + c\lambda'_I - d}{2c}\right]^2 - \frac{c\lambda'_R - d\lambda'_I}{c}, \tag{2.58}$$

$$Y'_{21} = r + jq, \quad Z'_{21} = c + jd. \tag{2.59}$$

Note that the center and the radius of each of these circles depend on all four elements of the S matrix of the core ($\mathbf{S}'$) through Y'_{21} or Z'_{21}.

Application of Y feedback [Figure 2.25(a)] to move from λ' onto the yellow strip in Figure 2.28 is shown in Figure 2.29 for a 40-nm NMOSFET at 235 GHz. In this example, the necessary susceptance B_Y is negative and can be implemented with inductive elements [26,27,34]. The frequency response of the resultant feedback amplifier (Design D1) is shown in Figure 2.30, together with two more amplifiers designed the same way at other frequencies (185 and 275 GHz). Since the goal point, λ, in Figure 2.29 is not very close to the highest gain point, G_{ma} is considerably lower than $G_{\text{max}}(1)$ at 235 GHz.

To aim for higher gain, Y and Z feedback can be applied successively. Figures 2.31 and 2.32 show two different designs, D2 and D3, with exactly the same starting point, λ'', and the goal point, λ, corresponding to $G_{\text{max}}(1.05)$. These two solutions were found by numerically solving the simultaneous equations, derivable from geometrical consideration on the λ-plane, for finding the values of B_Y and X_Z. Figure 2.33 shows the frequency responses of the two amplifiers. Both of the designs give $G_{\text{ma}} = G_{\text{max}}(1.05)$ at 235 GHz as they should, but the bandwidths are very different. Clearly, D2 is a superior design.

To explore the possibility of further improvement, we look at a less popular feedback configuration that uses a tapped transformer [31], shown in Figure 2.34. Such a feedback configuration was recently used to build a high-performance InP

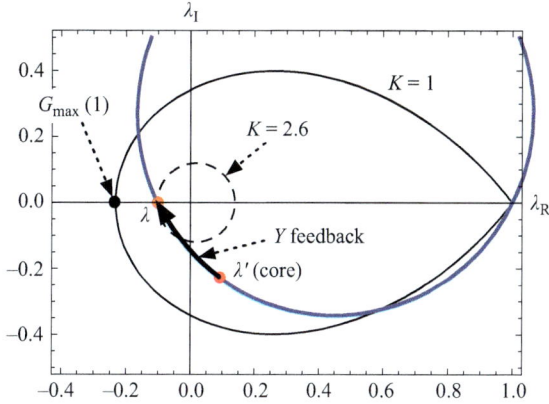

Figure 2.29 MAG-K chart for Design D1, in which Y feedback [Figure 2.25(a)] is applied with $B_Y = -3.2\,m\Omega$. The frequency is 235 GHz. The core is a 40-nm NMOSFET with $U = 1.63$ (2.05 dB) at 235 GHz and $f_{max} \simeq 292$ GHz. Copyright 2016 IEEE. Reproduced, with permission, from [37]

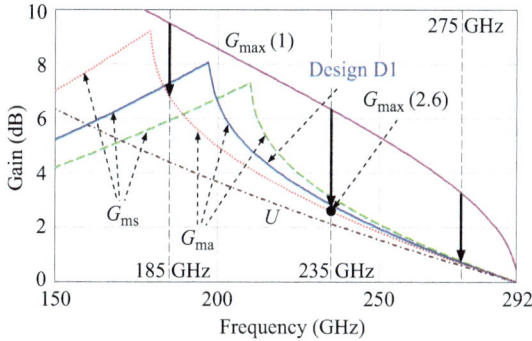

Figure 2.30 G_{ma} of Y-feedback amplifiers designed at 185, 235 (Design D1), and 275 GHz. B_Y is implemented with an inductor. The amplifiers are no longer unconditionally stable at frequencies below the inflection points, and therefore, G_{ms} is plotted there. Similar plots can be found in [26,34]. Copyright 2016 IEEE. Reproduced, with permission, from [37]

millimeter-wave amplifier [38]. If the transformer is ideal, the transformation formula is given by

$$\lambda = \frac{Y'_{12}(n-1) - Y'_{22}}{Y'_{21}(n-1) - Y'_{22}} = \frac{\lambda'(n-1) - Y'_{22}/Y'_{21}}{n-1 - Y'_{22}/Y'_{21}}. \tag{2.60}$$

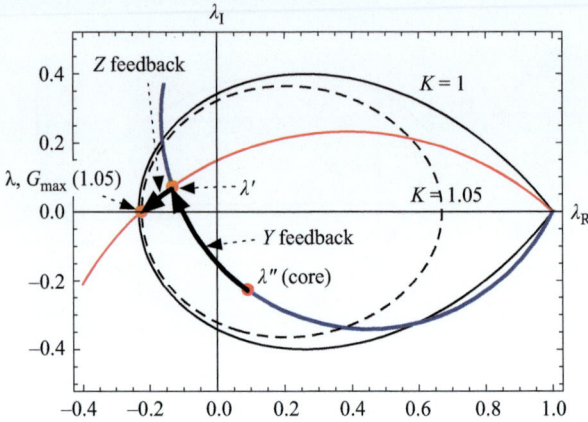

Figure 2.31   *MAG-K chart for Design D2, in which Y feedback ($B_Y = -3.9\,m\Omega$) and Z feedback ($X_Z = -62\,\Omega$) are applied. The transistor and the frequency are the same as in Figure 2.29. Copyright 2016 IEEE. Reproduced, with permission, from [37]*

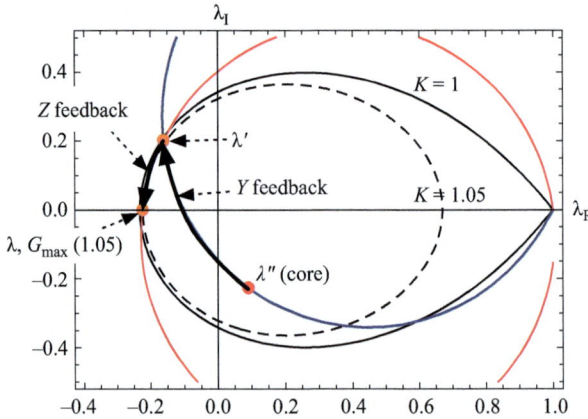

Figure 2.32   *MAG-K chart for Design D3, in which Y feedback ($B_Y = -5.0\,m\Omega$) and Z feedback ($X_Z = 495\,\Omega$) are applied. Note that the starting point λ'' and the goal λ are exactly the same as in Figure 2.31. Copyright 2016 IEEE. Reproduced, with permission, from [37]*

The locus is again a circle:

$$\left[\lambda_R - \frac{b\lambda'_R - a\lambda'_I + b}{2b}\right]^2 + \left[\lambda_I - \frac{a\lambda'_R + b\lambda'_I - a}{2b}\right]^2$$

$$= \left[\frac{b\lambda'_R - a\lambda'_I + b}{2b}\right]^2 + \left[\frac{a\lambda'_R + b\lambda'_I - a}{2b}\right]^2 - \frac{b\lambda'_R - a\lambda'_I}{b}, \tag{2.61}$$

$$Y'_{22}/Y'_{21} = a + jb. \tag{2.62}$$

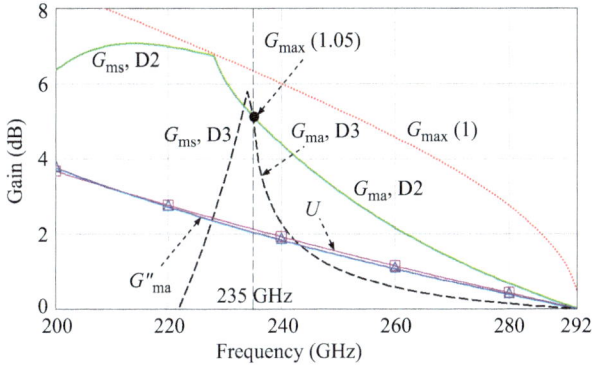

Figure 2.33   *Frequency dependence of U, core's G''_{ma}, $G_{max}(1)$, MAG and MSG of two amplifiers with Y–Z feedback (Designs D2 and D3), meant to achieve $G_{ma} = G_{max}(1.05)$ at 235 GHz. Negative B_Y in D2 and D3 are implemented with an inductor. Negative X_Z in D2 is implemented with a shorted transmission-line stub to secure a DC path. Positive X_Z in D3 is implemented with an inductor. Copyright 2016 IEEE. Reproduced, with permission, from [37]*

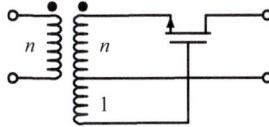

Figure 2.34   *Feedback with tapped ideal transformer. $n \to \infty$ corresponds to common-gate amplifier and $n \to 0$ corresponds to common-source amplifier. DC-cut capacitor needed for biasing is not shown. Copyright 2016 IEEE. Reproduced, with permission, from [37]*

When this feedback alone is applied to the same MOSFET as in the earlier examples, desired gain-boosting does not take place (Figure 2.35). This suggests that the transformer feedback should be combined with some other feedback. Figures 2.36 and 2.37 show such an example, where inductive Z feedback is followed by the transformer feedback. The inductive reactance may represent the inductance of a nonideal transformer. A notable feature of the resulting frequency response (Figure 2.38) is that G_{ma} is concave downward, unlike in Figure 2.33. This characteristic is highly desirable for wideband design.

We presented a graphical method of analyzing and designing gain-boosted near-f_{max} feedback amplifiers using the MAG-K chart (Figure 2.28). The chart helps the designer to graphically see possible design options. Specifically, a better feedback configuration than the one popularly used [26,27,34] could be found. It is also fairly easy to derive from the chart equations for finding necessary element values. Our method could facilitate near-f_{max} amplifier design.

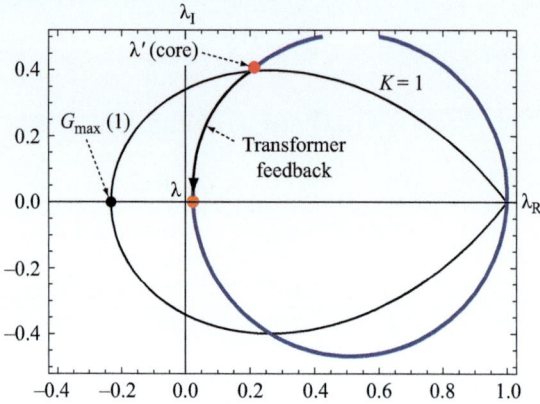

Figure 2.35   MAG-K chart for the tapped transformer feedback is shown in Figure 2.34. Starting point (red) is a common-gate MOSFET. The locus does not cross the yellow strip (Figure 2.28) when n is swept. Copyright 2016 IEEE. Reproduced, with permission, from [37]

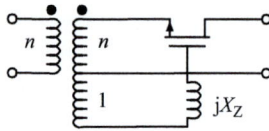

Figure 2.36   Z feedback applied to common-gate MOSFET, followed by tapped transformer feedback

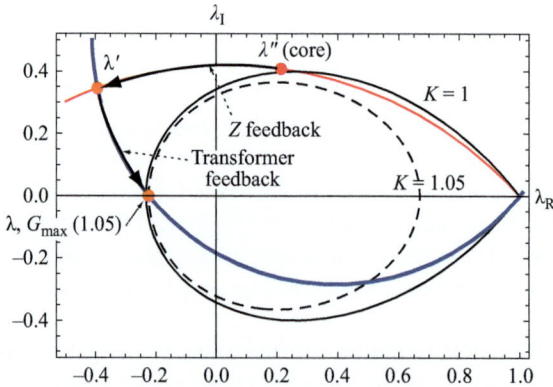

Figure 2.37   MAG-K chart for Figure 2.36. $X_Z = 69.4\,\Omega$ and $n = 1.545$ bring λ to the same goal point as Figures 2.31 and 2.32. Copyright 2016 IEEE. Reproduced, with permission, from [37]

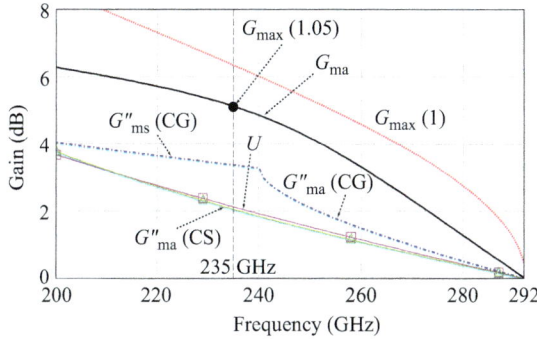

Figure 2.38 *Frequency dependence of U, core's G''_{ma} (common-source and common-gate), $G_{max}(1)$, and G_{ma} for the design in Figure 2.37. Copyright 2016 IEEE. Reproduced, with permission, from [37]*

2.3.5 Gain boosting using leaky tapped transformer

This section develops a theory on high-gain near-f_{max} feedback amplifier design using a tapped transformer with imperfect coupling [39]. It extends a design theory that assumed an ideal tapped transformer, reported recently. It is shown that by appropriately making corrections to feedback element values, the same gain-boosting as that obtained by the ideal treatment is possible.

Theoretical studies on design of gain-boosted near-f_{max} amplifiers using lossless reciprocal feedback have recently been reported [22,37,40,41]. All these papers are based on pioneering work by Singhakowinta and Boothroyd [31]. Among these papers, [37] and [31] looked at tapped transformer feedback. Reference [37] showed that by combining series–series and ideal tapped transformer feedback, gain boosting can be obtained over a wider bandwidth than is possible with the combination of series–series and shunt–shunt feedback, studied in [37,40,41]. However, the assumption of perfect coupling ($|k| = 1$, where k is transformer coupling coefficient) made it difficult to apply the theory [37] to practical design scenarios. We show how to deal with transformers with imperfect coupling ($|k| < 1$).

We investigate transformer feedback shown in Figure 2.40(a). A general representation of a feedback amplifier consisting of a core two-port (transistor) and an embedding four-port network is shown in Figure 2.27. Gain boosting is possible if the embedding four-port is appropriately designed [22,31], and that can be done using a graphical chart called a MAG-K chart [37]. An example is shown in Figure 2.39. It shows the MAG and the Rollett stability factor K on a complex plane spanned by $\lambda_R = \Re(\lambda)$ and $\lambda_I = \Im(\lambda)$, where [22,31]

$$\lambda = \frac{S_{12}}{S_{21}} = \frac{Y_{12}}{Y_{21}} = \frac{Z_{12}}{Z_{21}} \tag{2.63}$$

is complex inverse gain of a two-port (core or feedback amplifier). Analytic equations for constant-MAG and constant-K contours on the λ-plane are given in [37]. A core transistor's λ' is transformed into a new value λ by the four-port.

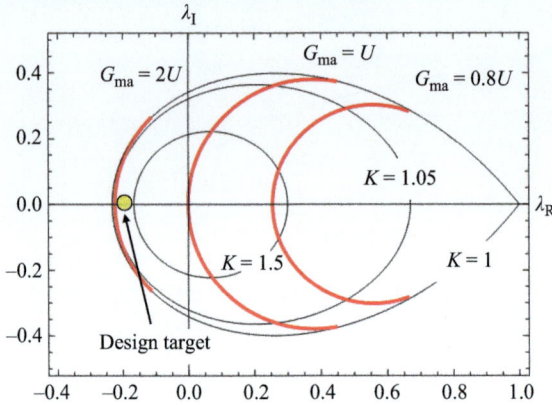

Figure 2.39   *MAG-K chart. The red arcs are constant-MAG contours. The black*
curves are constant-K contours. U is Mason's unilateral gain [29].
The yellow region gives the highest gain and good stability

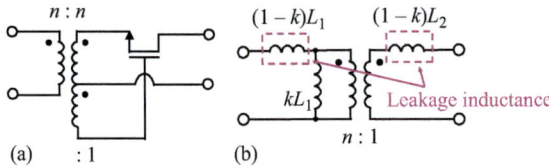

Figure 2.40   *(a) Tapped transformer feedback. (b) Equivalent circuit of a tapless*
imperfect (leaky) transformer with leakage inductance. k is the
coupling coefficient. L_1 and L_2 are self-inductances

If the transformer is ideal, the transformed λ is given by [31,37]

$$\lambda = \frac{Y'_{12}(n-1) - Y'_{22}}{Y'_{21}(n-1) - Y'_{22}}, \tag{2.64}$$

where Y'_{ij} are the Y-parameters of the core two-port (Figure 2.27). However, actual
transformers have some leakage and $|k| < 1$. A representation of a leaky transformer
is shown in Figure 2.40(b) [42].

To take leakage inductances into account, we move the leakage inductance on
the left-hand side as shown in Figure 2.41 [42]. Since a tapped transformer can be
composed of two tapless transformers, the tapped transformer feedback with leakage
inductances can be represented as shown in Figure 2.42. Transformed λ is given by

$$\lambda = \frac{Y'_{12}(N-1) - Y'_{22} - N\left(Z' + \frac{Z}{N}\right)\det \mathbf{Y'}}{Y'_{21}(N-1) - Y'_{22} - N\left(Z' + \frac{Z}{N}\right)\det \mathbf{Y'}}, \tag{2.65}$$

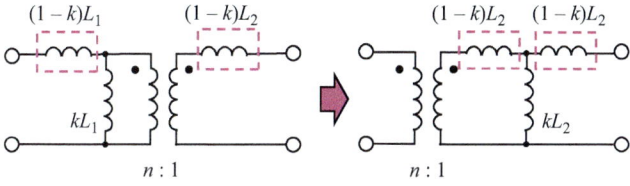

Figure 2.41   Leakage inductance on the left-hand side (in Figure 2.40(b)) can be moved to the right-hand side. The transformers are ideal

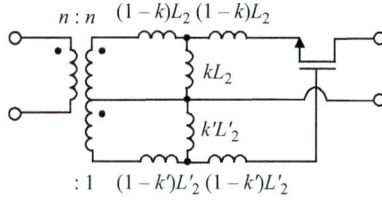

Figure 2.42   A transformer feedback with leakage inductances. k and L_2 are attributes of the upper transformer and k' and L'_2 are those of the lower transformer

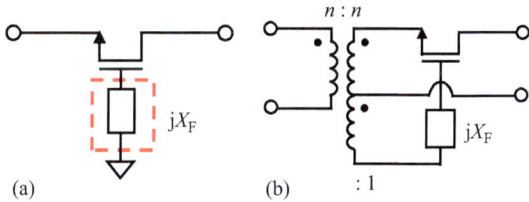

Figure 2.43   (a) Series–series or Z feedback and (b) combination of Z feedback and tapped transformer feedback

where $N = (k'/k)n$, $Z = j\omega(1 - k^2)L_2$, and $Z' = j\omega(1 - k'^2)L'_2$. $\mathbf{Y}'$ is the Y-matrix of the core two-port. When ideal transformer feedback (Figure 2.40(a)) is combined with Z feedback (Figure 2.43(a)) as shown in Figure 2.43(b),

$$\lambda = \frac{Y'_{12}(n - 1) - Y'_{22} - nX_F \det \mathbf{Y}'}{Y'_{21}(n - 1) - Y'_{22} - nX_F \det \mathbf{Y}'}. \tag{2.66}$$

By comparing (2.65) and (2.66), we see that N can be regarded as the effective turns ratio and that $(Z + (Z'/N))$ works as a Z-feedback element X_F. Therefore, when Z feedback is combined with imperfect tapped transformer feedback, we need to recalculate Z feedback element value, X_F, considering k, k', L_2, and L'_2.

To check the validity of the theoretical prediction, we simulated a feedback amplifier that uses Z feedback and tapped transformer feedback. We used the same

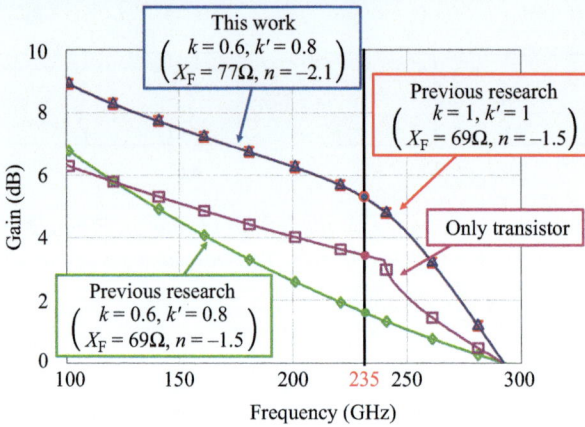

This work
$$\left(\begin{array}{c} k = 0.6, k' = 0.8 \\ X_F = 77\Omega, n = -2.1 \end{array} \right)$$

Previous research
$$\left(\begin{array}{c} k = 1, k' = 1 \\ X_F = 69\Omega, n = -1.5 \end{array} \right)$$

Only transistor

Previous research
$$\left(\begin{array}{c} k = 0.6, k' = 0.8 \\ X_F = 69\Omega, n = -1.5 \end{array} \right)$$

Figure 2.44   Simulated MAG of a common-gate NMOSFET (pink), an amplifier with ideal transformer and Z feedback [37] (red), an amplifier with leaky transformer and the same Z feedback (green), and an amplifier with leaky transformer and recalculated Z feedback (blue). The design frequency is 235 GHz

40-nm NMOS transistor model as that used in [37]. The design frequency is 235 GHz. Simulation results are shown in Figure 2.44. By appropriately making a correction to X_F depending on k, k', L_2, and L'_2, exactly the same performance as that from the ideal treatment [37] can be obtained.

In conclusion, we have derived an equation that can be used to design gain-boosted amplifier with imperfect tapped transformer feedback. It is more useful than the earlier work [37] that assumed that the transformer was ideal.

2.4   Compact layout techniques [43], [53]

2.4.1   "Fishbone" layout for single-ended amplifiers

Many-stage amplifiers for terahertz frequencies tend to occupy a large area since interstage matching networks consist typically of several passive devices that are much larger than MOSFETs. To realize cost-effective chips, area reduction is important. In order to reduce the area of amplifiers, we proposed the "fishbone layout" [43], shown in Figure 2.45. In this technique, transmission line stubs used in matching networks are arranged regularly at narrow spacings, and the TLs themselves are designed to be narrow, thereby reducing the footprint. A chip micrograph of an eight-stage 160-GHz amplifier with fishbone layout is shown in Figure 2.46. The core size is as small as $190 \times 123\ \mu m^2$. Measured small-signal characteristics of the 160-GHz amplifier are shown in Figure 2.47. The peak gain is 14.9 dB at 160 GHz with the power consumption of 117 mW at the supply voltage of 0.9 V.

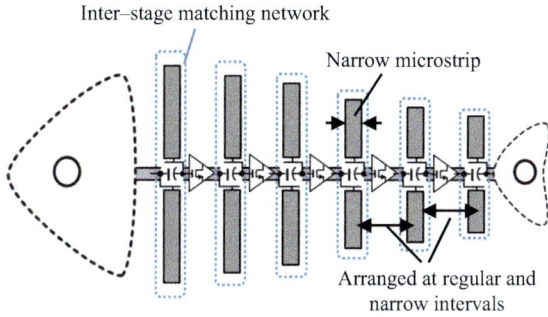

Figure 2.45 Concept of the area-saving "fishbone layout". Copyright 2016 IEICE. Reproduced, with permission, from [53]

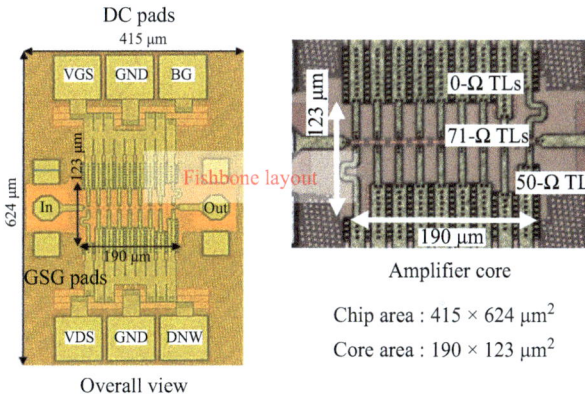

Figure 2.46 Chip micrographs of an eight-stage 160-GHz amplifier. Shunt stubs are made of narrow 71-Ω TLs. Wider 50-Ω TLs are used for signal I/O

2.4.2 Extension of "fishbone" for differential amplifiers

A differential amplifier, which is well suited for driving differential signals, provides a better common-mode rejection ratio and power supply rejection ratio through its insensitivity to noise and interference; it is also used as a driver amplifier for element circuits such as an unbalance mixer in transceiver integration. A compact differential amplifier could be built with shunt stubs like in the proposed fishbone-layout technique instead of using transformers in an interstage matching network. However, a differential amplifier needs twice the number of shunt stubs compared with single-ended amplifier because of the differential signal lines. If the fishbone layout could be adapted to a differential amplifier, the shunt stub would need to cross the signal line. This would increase the matching network area and risk adverse effects on the amplifier characteristics caused by crosstalk between the signal and bias lines. Therefore, the fishbone layout technique cannot be easily adapted to a differential

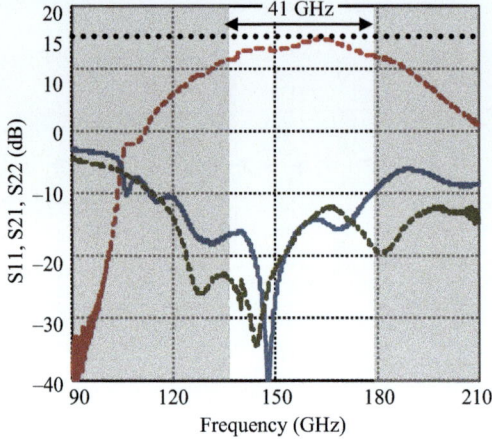

Figure 2.47   Measured S parameters of the amplifier in Figure 2.46

amplifier. In order to achieve a small footprint for a differential amplifier, one needs further ingenuity. Here we demonstrate a compact wideband five-stage differential amplifier designed for a 40-nm CMOS process. A small core area of $201 \times 284\,\mu m^2$, not including that of the input/output matching networks, was achieved by using a refined "fishbone" layout technique.

2.4.3   Design

To design a compact differential amplifier in which the interstage matching network needs four shunt stubs because of differential signal lines, a special T-junction composed of tenth metal layers was put between the amplifier stage and the interstage capacitor. Two pairs of shunt stubs grew out from the north and south edges of T-junction against each signal line in the interstage network. The shunt stubs were arranged without crossing the signal lines with a regular and narrow spacing (Figure 2.48(a)). This "millipede" layout that makes an interstage matching network compact (Figure 2.48(b)) can provide a small footprint like that of the fishbone layout for a single-ended amplifier.

2.4.3.1   Transmission lines

The compact differential amplifier was designed for the TSMC 40-nm 1P10M CMOS GP process. Its back end consisted of ten copper layers and a top aluminum redistribution layer (RDL). Grounded coplanar waveguide (GCPW) TLs with a characteristic impedance of Z_0 of 50 Ω (50-Ω GCPW TL) and 71 Ω (71-Ω GCPW TL) were designed with an electromagnetic (EM) simulation by ANSYS HFSS. A 50-Ω GCPW TL was used for connecting the signal pads and shunt stub of the input/output matching (Figure 2.49). Its signal line was composed of an RDL layer with a width of 9.0 μm. Ground (GND) walls composed of the sixth to tenth metal layers with a width of 2.7 μm were placed on both sides of the signal line at the distance of 7.2 μm. A 71-Ω

V_{DS} V_{GS} V_{DS} V_{GS}

Interstage matching networks

(a)

Capacitor and T-lines (*somite*)

Narrow TL stub (*leg*)

Arranged at regular and narrow intervals

Interstage matching network

(b)

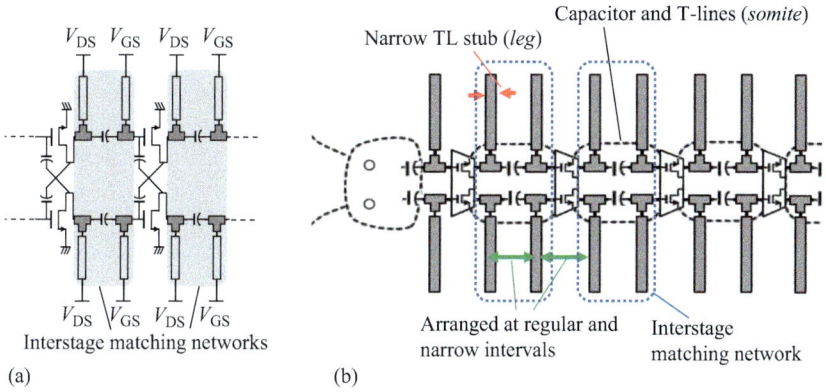

Figure 2.48 *(a) Schematic layout of proposed compact interstage matching networks and (b) illustration of the "millipede" layout concept. Copyright 2016 IEICE. Reproduced, with permission, from [53]*

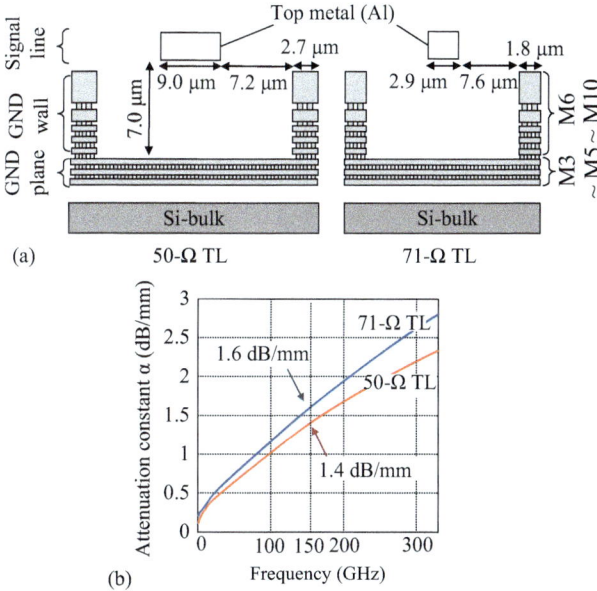

Top metal (Al)

Signal line

GND plane   GND wall

7.0 µm

9.0 µm   7.2 µm   2.7 µm

2.9 µm   7.6 µm   1.8 µm

M6   M3

M5 ~ M10

Si-bulk

Si-bulk

(a)   50-Ω TL   71-Ω TL

71-Ω TL

1.6 dB/mm

50-Ω TL

1.4 dB/mm

Attenuation constant α (dB/mm)

Frequency (GHz)

(b)

Figure 2.49 *(a) Cross section of GCPW TLs and (b) EM simulation results of the attenuation constant, α, of the 50- and 71-Ω GCPW TLs as a function of frequency. Copyright 2016 IEICE. Reproduced, with permission, from [53]*

GCPW TL designed for a rat-race balun was also used for the shunt stubs of the interstage matching network. The RDL-layer signal line was 2.9 µm wide, and the line was 1.8 µm wide (Figure 2.49(a)). The third to fifth metal GND wall placed at a distance of 7.6 µm from the signal layers was meshed and stitched together with vias

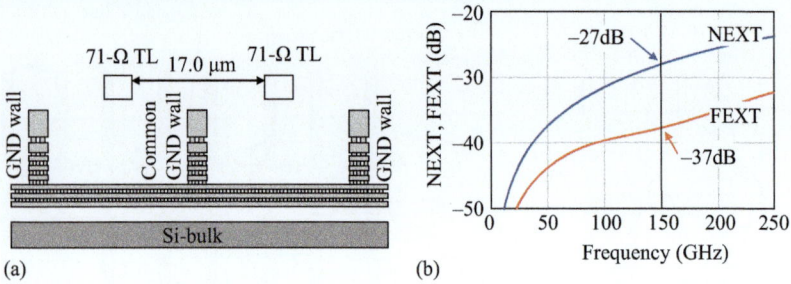

Figure 2.50   *(a) Cross section of two 71-Ω GCPW TLs with the common GND wall and (b) EM simulation results of near-end (NEXT) and far-end (FEXT) crosstalk as a function of frequency. Copyright 2016 IEICE. Reproduced, with permission, from [53]*

to form the GND plane. The EM simulation predicted the attenuation constants, α, of the 50- and 71-Ω GCPW TLs to be 1.4 and 1.6 dB/mm at 150 GHz, respectively. Although α was slightly larger, the narrow 71-Ω GCPW TLs for the shunt stubs provided compactness.

2.4.3.2   Crosstalk between shunt stubs

The shunt stubs in the interstage networks were arranged to be adjacent across a common GND wall, and the space between the lines was 17 μm (Figure 2.50(a)). The near-end and far-end crosstalk in the EM simulation were −27 and −37 dB at 150 GHz, respectively (Figure 2.50(b)). The shunt stub also played the role of a power line, and the crosstalk voltage-ratio was below 0.2%. In particular, the crosstalk voltage was below 2 mV when 0.94 V was applied to a stub, and the variation was small. This indicates that the cross-coupling between stubs was negligible. In addition, the common GND wall between signal lines was composed of individual metal layers. Using this layout, metal fill density was high enough for processing including chemical mechanical polishing without using extra dummy metals.

2.4.3.3   Capacitive cross-coupling technique

A capacitive cross-coupling technique was used for the design of a common-source differential amplifier (Figure 2.48(a)) [44,45]. This technique neutralizes the parasitic of the gate-to-drain capacitance, C_{gd}, and leads to an improvement in power gain and reverse isolation. Figure 2.51 shows the simulation results for the MAG, MSG, and stability factor (K) depending on the neutralization capacitance of the cross-coupling capacitor. The applied gate voltage and drain voltage were 0.8 and 0.9 V, respectively. The frequency was 140 GHz. MAG and MSG are defined in the following equations. In this case, $K > 1$. Therefore, MAG has a conical depression shape against MSG. This result shows that MAG can be improved by about 20% by using an appropriate capacitance depending on the total gate width of the MOSFETs. On the basis of this result, we decided the neutralization capacitance for the differential amplifier.

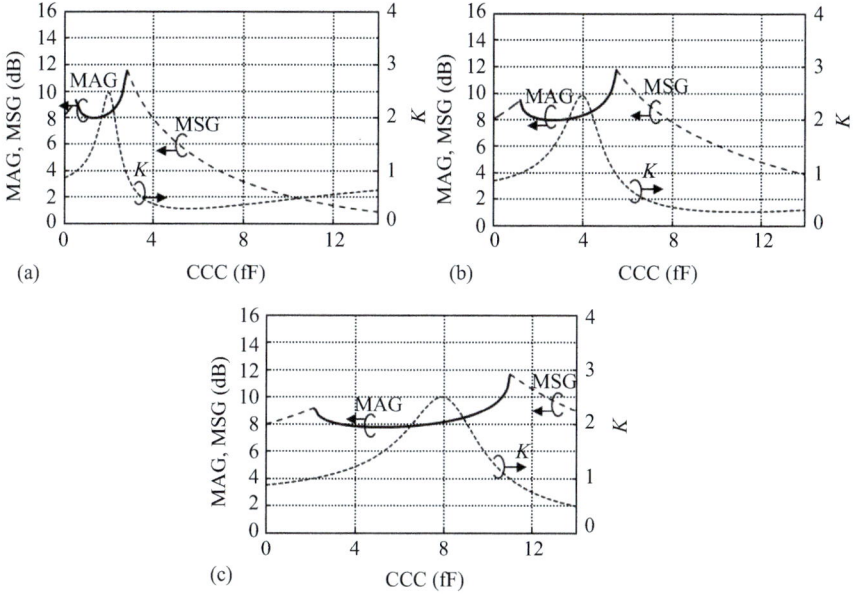

Figure 2.51 *Simulation results of MAG, MSG and K of a cross-coupling amplifier stage as a function of neutralization capacitance at 140 GHz (total gate widths of MOSFET are (a) 8, (b) 16, and (c) 32 μm). Copyright 2016 IEICE. Reproduced, with permission, from [53]*

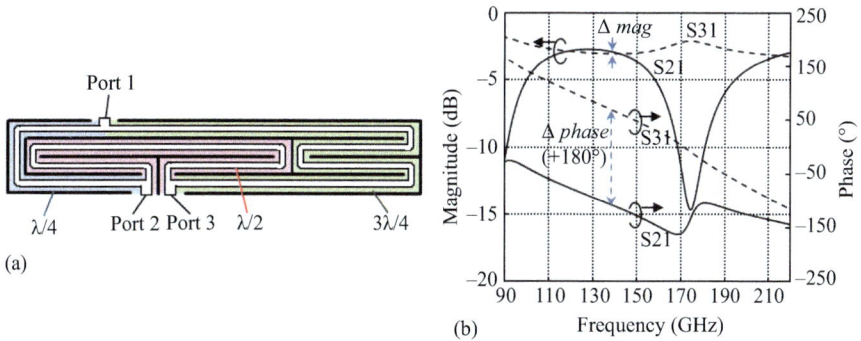

Figure 2.52 *(a) Schematic layout of the rat-race balun and (b) EM simulation result of S_{21}, S_{31} and phase as a function of frequency. Copyright 2016 IEICE. Reproduced, with permission, from [53]*

2.4.3.4 Rat-race balun

A rat-race balun with the 71-Ω GCPW TL was put at the input/output of the amplifier to convert the single-ended signal into a differential one. The balun was designed by bending the TLs of $\lambda/4$, $\lambda/2$, $3\lambda/4$ (λ is the wavelength of the 71-Ω GCPW TL, 310 μm) by using the common GND wall described above (Figure 2.52(a)); its size

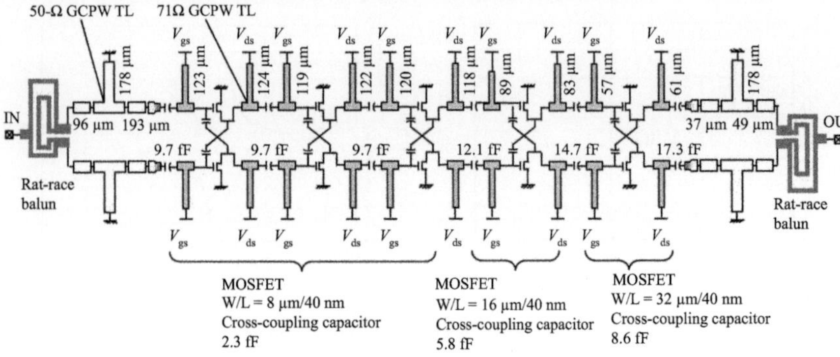

Figure 2.53   Circuit schematic of the proposed differential amplifier. Copyright 2016 IEICE. Reproduced, with permission, from [53]

was $80 \times 470\,\mu m^2$. Figure 2.52(b) shows the EM simulation results for S_{21}, S_{31}, and phase. The insertion loss was about 3 dB, and the amplitude error, Δmag, and phase error, Δphase, between balanced terminals were 0.23 dB and $2.2°$ at 140 GHz.

2.4.3.5   Five-stage differential amplifier

Figure 2.53 shows the circuit schematic of the amplifier. The differential amplifier with capacitive cross-coupling technique consists of five common-source stages. To improve gain and power, the first, intermediate, and last stages have MOSFETs with gate widths of 8, 16, 32 μm, respectively. The neutralization capacitances of the cross-coupling capacitor of each stage were chosen for obtaining a high gain on the basis of the simulation results shown in Figure 2.51. The shunt stubs that have common GND walls are regularly arranged with a spacing of 17 μm. The amplifier was designed as an IF amplifier, and it was planned to be connected with another element circuit of which the matching circuit included a 50-Ω GCPW TL. Therefore, an input/output matching circuit with a 50-Ω GCPW TL was also designed. The T-junction consisted of tenth metal layers connects between the MOSFET and matching capacitor. Two pairs of shunts stub grew out from the north and south edges of the T-junction in an interstage network. The lengths of the TL stubs and the sizes of rotative metal–oxide–metal (RTMOM) capacitors were determined in a nonparametric optimization process that took into account the models of the MOSFET, RTMOM, TLs, and RF pad [24]. The far ends of the shunt stubs were terminated by wideband decoupling power lines (0-Ω TLs) with lengths exceeding 140 μm [46].

2.4.4   Results and discussion

Figure 2.54 shows a microphotograph of the fabricated amplifier. The input and output GSG pads were designed to accommodate 75-μm-pitch GSG probes. The chip area is $945 \times 842\,\mu m^2$. The amplifier core without input/output matching network, which was designed for connecting to the other circuits, is $201 \times 284\,\mu m^2$. The small-signal measurement was performed using OML WR-8 extenders in 90–140 GHz and WR-5 extenders in 140–220 GHz with a Keysight Technologies PNA-X5247A vector

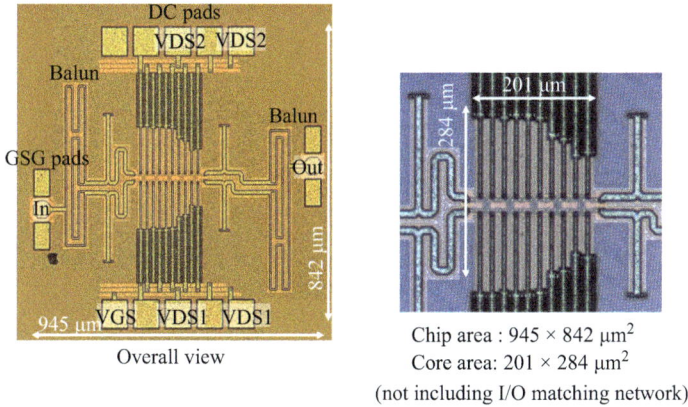

Overall view

Chip area : 945 × 842 μm²
Core area: 201 × 284 μm²
(not including I/O matching network)

*Figure 2.54   Chip microphotograph of the five-stage differential amplifier.
Copyright 2016 IEICE. Reproduced, with permission, from [53]*

network analyzer (VNA) and an N5242A test set. GGB Model 140 and Model 220 Picoprobes were used for the measurements at 90–140 GHz and at 140–220 GHz, respectively. The thru–reflect–line calibration was performed on a GGB CS-15 calibration substrate. Figure 2.55 shows the results of the small-signal measurement and simulation of the compact five-stage amplifier. The peak gain was 20 dB at 148 GHz (Figure 2.55(c)). The 3-dB bandwidth was 22 GHz, from 130 to 152 GHz (141 GHz center). Although the gain and bandwidth were slightly shifted, the measurements were close to the simulation results. The measured and simulated results of S_{11} and S_{22} were also very close. The power consumption was 75 mW. The DC supply voltage was 0.94 V. These results indicate that a high-gain and wideband amplifier was successfully realized.

The NF measurements were performed using an ELVA-1 ISSN-06 noise source, a Keysight Technologies N8975A NF analyzer, and a Millitec MSH-06-2 harmonic mixer driven by a Keysight Technologies E8244A signal generator with an 83557A source module. The NF measurements are shown in Figure 2.56. The minimum NF was 8.5 dB at 135 GHz. The large signal measurements are shown in Figure 2.57. The input power from a VNA extender was varied using a level set attenuator, and the output was measured using a VDI PM-5 power sensor. The power at the probe tip was calibrated with the measured loss of the WR-5 S-bend waveguide and probe, which was about 4.5 dB at 110–170 GHz. The large signal gain in the linear region was 18 dB at 138 GHz, which was almost the same as the small signal gain shown in Figure 2.55(b). The 1-dB compression point (P_{1dB}) was 1.7 dB m. Although the correct amplifier saturated power (P_{sat}) and power-added efficiency was not estimated, we concluded that they were slightly larger than 4.5 dB m and 3.6%, respectively.

Table 2.1 summarizes the amplifier performance in comparison with prior work. The amplifier had a wide bandwidth and high gain and was compact. The gain–bandwidth product (GBW) was estimated to be 220 GHz from the gain and 3-dB bandwidth. Although the GBW was slightly lower than that in [43], this was caused

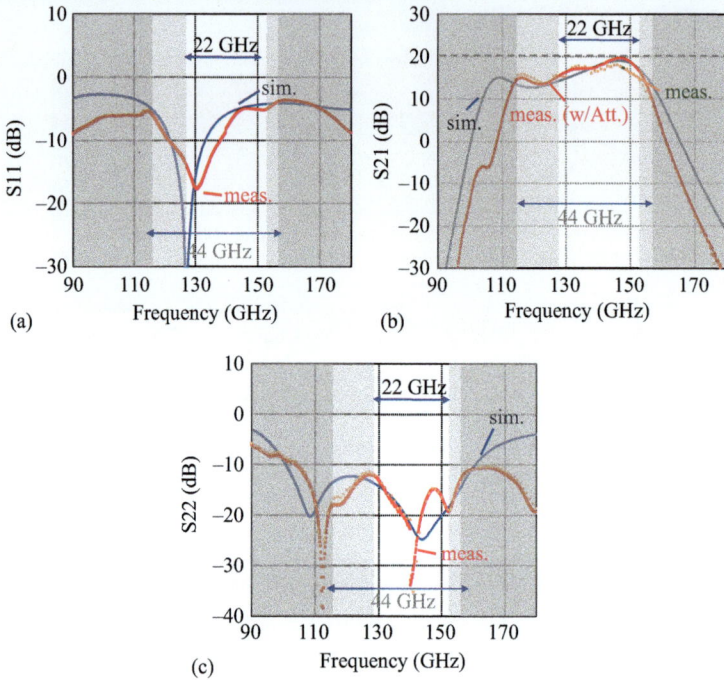

(a)

(b)

(c)

Figure 2.55   *Measured and simulated results of five-stage differential amplifier for*
a small signal ((a) S_{11}, (b) S_{21}, and (c) S_{22}). Copyright 2016 IEICE.
Reproduced, with permission, from [53]

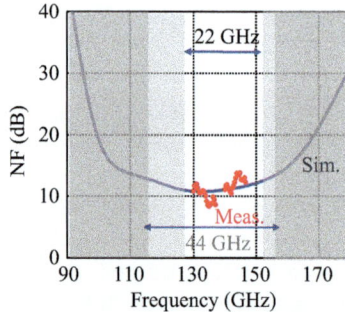

Figure 2.56   *Noise figure of five-stage differential amplifier as a function of*
frequency. Copyright 2016 IEICE. Reproduced, with permission,
from [53]

by the difference in the number of amplifier stages. Among the differential amplifiers
that have the same advantages regarding noise and interference, the series length per
stage of the current amplifier is the shortest. These results indicate that the amplifier
embodying the "millipede" layout technique can reduce costs because it has a small
footprint without sacrificing performance.

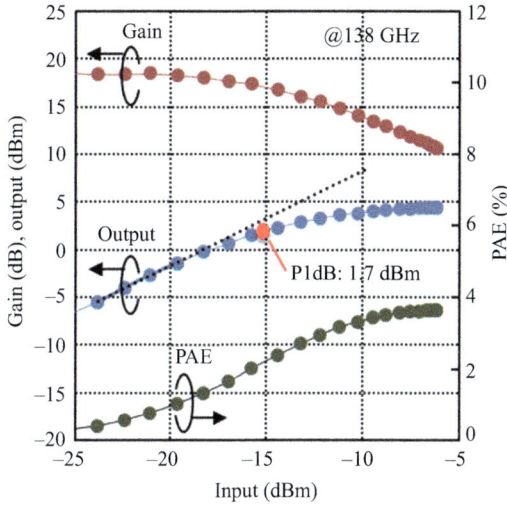

Figure 2.57 Measured results of five-stage differential amplifier for a large signal at 138 GHz. Copyright 2016 IEICE. Reproduced, with permission, from [53]

Table 2.1 Performance summary and comparison with prior work

Reference	Technology	No. of stages/ Architecture	Frequency (GHz)	3-dB BW (GHz)	Core size (μm²)	Gain (dB)	GBW (GHz)
[47]	40 nm CMOS	9/Single-ended	210	13	130 × 103	10.5	43.5
[48]	65 nm CMOS	4/Single-ended	140	15	560 × 347	15	78.7
[49]	65 nm CMOS	5/Single-ended	170	14*	540 × 430*	19	118.5
This [43]	40 nm CMOS	8/Single-ended	160	41	190 × 120	14.9	241.8
[50]	65 nm CMOS	3/Differential	144	3*	480 × 100	20.6	40.7
[51]	65 nm CMOS	5/Differential	200	5*	680 × 85	8.1	12.7
[52]	65 nm CMOS	4/Differential	147	13*	580 × 200*	7.1	29.4
This [53]	40 nm CMOS	5/Differential	141	22	201 × 284†	20.0	220

*Graphical estimate.
†Not including input/output matching networks.

2.4.5 Conclusion

A 141-GHz amplifier was designed for the 40-nm CMOS process. The core was kept reasonably compact $201 \times 284\,\mu m^2$ in spite of the five-stage differential amplifier by using a layout avoiding the coupling effect of adjacent shunt stub lines. The peak gain and 3-dB bandwidth were 20 dB at 148 and 22 GHz, respectively. The power consumption was 75 mW from a 0.94-V power supply. The minimum NF was 8.5 dB at 135 GHz. The P1dB was 1.7 dB m at 138 GHz. The proposed "millipede" layout technique enables an amplifier with a small footprint to realize without sacrificing performance.

References

[1]   W. W. Gärtner, "Maximum available power gain of linear four-poles," *IRE Transactions on Circuit Theory*, vol. 5, no. 4, pp. 375–376, 1956.

[2]   A. P. Stern, "Stability and power gain of tuned transistor amplifiers," *Proceedings of the IRE*, vol. 45, no. 3, pp. 335–343, 1957.

[3]   E. F. Bolinder, "Survey of some properties of linear networks," *IRE Transactions on Circuit Theory*, vol. 4, no. 3, pp. 70–78, 1957; Corrections: *IRE Transactions on Circuit Theory*, vol. 5, no. 2, p. 139, 1958.

[4]   M. A. Karp, "Power gain and stability," *IRE Transactions on Circuit Theory*, vol. 4, no. 4, pp. 339–340, 1957.

[5]   S. Voinigescu, *High-Frequency Integrated Circuits*, Cambridge University Press, Cambridge, 2013.

[6]   G. Gonzalez, *Microwave Transistor Amplifiers: Analysis and Design*, 2nd edition, Prentice Hall, Upper Saddle River, NJ, 1996.

[7]   R. Ludwig, G. Bogdanov, *RF Circuit Design: Theory and Applications*, 2nd edition, Pearson Prentice Hall, Upper Saddle River, NJ, 2009.

[8]   R. E. Collin, *Foundations for Microwave Engineering*, 2nd edition, Wiley-Interscience, New York, 2001.

[9]   R. Mavaddat, *Network Scattering Parameters*, World Scientific, Singapore, 1996.

[10]   Z. Wang, P.-Y. Chiang, P. Nazari, C.-C. Wang, Z. Chen, P. Heydari, "A 210 GHz fully integrated differential transceiver with fundamental-frequency VCO in 32 nm SOI CMOS," *International Solid-State Circuits Conference*, pp. 136–137, Feb. 2013.

[11]   M. Fujishima, M. Motoyoshi, K. Katayama, K. Takano, N. Ono, R. Fujimoto, "98 mW 10 Gbps wireless transceiver chipset with D-band CMOS circuits," *IEEE Journal of Solid-State Circuits*, vol. 48, no. 10, pp. 2273–2284, 2013.

[12]   M. L. Edwards, S. Cheng, J. H. Sinsky, "A deterministic approach for designing conditionally stable amplifiers," *IEEE Transactions on Microwave Theory and Techniques.*, vol. 43, no. 7, pp. 1567–1575, 1995.

[13]   L. I. Babak, "Comments on 'A deterministic approach for designing conditionally stable amplifiers'," *IEEE Transactions on Microwave Theory and Techniques*, vol. 47, no. 2, pp. 250–251, 1999.

[14]   S. Mizukusa, K. Takano, K. Katayama, S. Amakawa, T. Yoshida, M. Fujishima, "Analytical design of small-signal amplifier with maximum gain in conditionally stable region," *Asia-Pacific Microwave Conference*, pp. 774–776, Nov. 2014.

[15]   S. Roberts, "Conjugate-image impedances," *Proceedings of the IRE*, vol. 34, no. 4, pp. 198P–204P, 1946.

[16]   J. Choma, W. K. Chen, *Feedback Networks: Theory and Circuit Applications*, World Scientific, Singapore, 2007.

[17]   G. V. Petrov, "Analysis and design of microwave-linear transistor amplifiers," *International Journal of Electronics*, vol. 56, no. 5, pp. 641–647, 1984.

[18] B. M. Albinsson, "A graphic design method for matched low-noise amplifiers," *IEEE Transactions on Microwave Theory and Techniques*, vol. 38, no. 2, pp. 118–122, 1990.

[19] K. W. Eccleston, "Optimum design of small-signal microwave amplifiers with specified stability safety margin," *Asia-Pacific Microwave Conference*, pp. 863–866, Dec. 2000.

[20] R. S. Engelbrecht, K. Kurokawa, "A wide-band low noise L-band balanced transistor amplifier," *Proceedings of the IEEE*, vol. 53, no. 3, pp. 237–247, 1965.

[21] M. Fujishima. "Terahertz CMOS electronics for future mobile applications," *ECS Transactions*, vol. 61, no. 6, pp. 43–50, 2014.

[22] S. Amakawa, "Theory of gain and stability of small-signal amplifiers with lossless reciprocal feedback," *Asia-Pacific Microwave Conference*, pp. 1184–1186, Nov. 2014.

[23] M. Fujishima, S. Amakawa, K. Takano, K. Katayama, and T. Yoshida, "Terahertz CMOS design for low-power and high-speed wireless communication," *IEICE Transactions on Electronics*, vol. E98-C, no. 12, pp. 1091–1104, 2015.

[24] K. Katayama, M. Motoyoshi, K. Takano, C. Y. Li, S. Amakawa, M. Fujishima, "E-band 65nm CMOS low-noise amplifier design using gain-boost technique," *IEICE Transactions on Electronics*, vol. E97-C, no. 6, pp. 476–485, 2014.

[25] H. T. Friis, "Noise figure of radio receivers," *Proceedings of the IRE*, vol. 32, no. 7, pp. 419–422, 1944.

[26] O. Momeni, E. Afshari, "A high gain 107 GHz amplifier in 130 nm CMOS," *Custom Integrated Circuits Conference*, pp. 1–4, Sep. 2011.

[27] O. Momeni, "A 260 GHz amplifier with 9.2 dB gain and −3.9 dB m saturated power in 65 nm CMOS," *International Solid-State Circuits Conference*, pp. 140–141, Feb. 2013.

[28] P.-O. Leine, "On the power gain of unilaterized active networks," *IRE Transactions on Circuit Theory*, vol. 8, no. 3, pp. 357–358, 1961.

[29] S. J. Mason, "Power gain in feedback amplifier," *IRE Transactions on Circuit Theory*, vol. 1, no. 2, pp. 20–25, 1954.

[30] G. D. Vendelin, A. M. Pavio, U. L. Rohde, *Microwave Circuit Design Using Linear and Nonlinear Techniques*, 2nd edition, Wiley, New York, 2005.

[31] A. Singhakowinta, A. R. Boothroyd, "Gain capability of two-port amplifiers," *International Journal of Electronics*, vol. 21, no. 6, pp. 549–560, 1966.

[32] A. Singhakowinta, A. R. Boothroyd, "On linear two-port amplifiers," *IEEE Transactions on Circuit Theory*, vol. 11, no. 1, p. 169, 1964.

[33] M. S. Gupta, "Power gain in feedback amplifiers, a classic revisited," *IEEE Transactions on Microwave Theory and Techniques*, vol. 40, no. 5, pp. 864–879, 1992.

[34] S. Moghadami, J. Isaac, S. Ardalan, "A 0.2–0.3 THz CMOS amplifier with tunable neutralization technique," *IEEE Transactions on Terahertz Science and Technology*, vol. 5, no. 6, pp. 1088–1093, 2015.

[35]   M. Vehovec, L. Houselander, R. Spence, "On oscillator design for maximum power," *IEEE Transactions on Circuit Theory*, vol. 15, no. 3, pp. 281–283, 1968.

[36]   P. D. van der Puije, R. Spence, "Power gain sensitivity of active devices," *International Journal of Electronics*, vol. 26, no. 1, pp. 49–65, 1969.

[37]   S. Amakawa, Y. Ito, "Graphical approach to analysis and design of gain-boosted near-f_{max} feedback amplifiers," *European Microwave Conference*, pp. 1039–1042, Oct. 2016.

[38]   M. Sato, S. Shiba, H. Matsumura, *et al.*, "93–133 GHz band InP high-electron-mobility transistor amplifier with gain-enhanced topology," *Japanese Journal of Applied Physics*, vol. 52, p. 04CF03, 2013.

[39]   S. Yasuda, S. Amakawa, "Gain-boosted feedback amplifier design using leaky tapped transformer," *Thailand-Japan Microwave*, Jun. 2017.

[40]   Z. Wang, P. Heydari, "A study of operating condition and design methods to achieve the upper limit of power gain in amplifiers at near-f_{max} frequencies," *IEEE Transactions on Circuits and Systems I: Regular Papers*, vol. 62, no. 2, pp. 261–271, 2017.

[41]   H. Bameri, O. Momeni, "A high-gain mm-wave amplifier design: an analytical approach to power gain boosting," *IEEE Journal of Solid-State Circuits*, vol. 52, no. 2, pp. 357–370, 2017.

[42]   I. Bahl, *Lumped Elements for RF and Microwave Circuits*, Artech House, Norwood, MA, pp. 317–324, 2003.

[43]   S. Hara, K. Katayama, K. Takano, *et al.*, "Compact 160-GHz amplifier with 15-dB peak gain and 41-GHz 3-dB bandwidth," *Radio Frequency Integrated Circuits Symposium*, pp. 7–10, May 2015.

[44]   W. L. Chan, J. R. Long, M. Spirito, J. J. Pekarik, "A 60 GHz-Band 1V 11.5 dB m power amplifier with 11% PAE in 65 nm CMOS," *International Solid-State Circuits Conference*, pp. 380–381, Feb. 2009.

[45]   H, Asada, K. Matsushita, K. Bunsen, K. Okada, A. Matsuzawa, "A 60GHz CMOS power amplifier using capacitive cross-coupling neutralization with 16% PAE," *European Microwave Conference*, pp. 1115–1118, Oct. 2011.

[46]   S. Amakawa, R. Goda, K. Katayama, K. Takano, T. Yoshida, M. Fujishima, "Wideband CMOS decoupling power line for millimeter-wave applications," *IEEE MTT-S International Microwave Symposium*, pp. 1–4, May 2015.

[47]   C.-L. Ko, C.-H. Li, C.-N. Kuo, M.-C.Kuo, D.-C. Chang, "A 210-GHz amplifier in 40-nm digital CMOS technology," *IEEE Transactions on Microwave Theory and Techniques*, vol. 61, no. 6, pp. 2438–2446, 2013.

[48]   Z.-M. Tsai, H.-C. Liao, Y.-H. Hsiao, *et al.*, "A 1.2V broadband D-band amplifier with 13.2-dB m output power in standard 65-nm CMOS," *IEEE MTT-S International Microwave Symposium*, pp. 1–3, Jun. 2012.

[49]   P.-H. Chen, J.-C. Kao, T.-L. Yu, *et al.*, "A 110–180 GHz broadband amplifier in 65-nm CMOS process," *IEEE MTT-S International Microwave Symposium*, pp. 1–3, Jun. 2013.

[50] Z. Xu, Q. J. Gu, M.-C. F. Chang, "A three stage, fully differential 128–157 GHz CMOS amplifier with wide band matching," *IEEE Microwave and Wireless Components Letters*, vol. 21, no. 10, pp. 550–552, 2011.

[51] Z. Xu, Q. J. Gu, M.-C. F. Chang, "200 GHz CMOS amplifier working close to device f_T," *Electronics Letters*, vol. 47, no. 11, pp. 639–641, 2011.

[52] C.-H. Li, C.-W. Lai, C.-N. Kuo, "A 147 GHz fully differential D-band amplifier design in 65-nm CMOS," *Asia Pacific Microwave Conference*, pp. 691–693, Nov. 2013.

[53] S. Hara, K. Katayama, K. Takano, *et al.*, "Compact 141-GHz differential amplifier with 20-dB peak gain and 22-GHz 3-dB bandwidth," *IEICE Transactions on Electronics*, vol. E99-C, no. 10, pp. 1156–1163, 2016.

Chapter 3
Physical design techniques for RF CMOS

Even in the case of a terahertz circuit design, circuit simulators are still used in circuit design. However, design environments and techniques are not as well established as for radio-frequency (RF) complementary metal-oxide-semiconductor (CMOS) circuits. Thus, various preparations are required before using the circuit simulator to design ultrahigh-frequency circuits operating at millimeter wave (mmw) or terahertz (Figure 3.1). Although this preparation occasionally occupies a lot of time of work, it is no exaggeration to say that the success or failure of circuit design is determined by the quality of this preparation. This section reviews recent progress made by the authors in terahertz CMOS design including device characterization and modeling techniques.

3.1 Physical design [1]

3.1.1 Bond-based design

At mmw frequencies, it is imperative that layout parasitics including resistances, capacitances, and inductances (both self and mutual) be appropriately taken into consideration. However, ordinary layout parasitic extraction (LPE) tools used for chip design do not extract inductances. Electromagnetic (EM) simulation should, in principle, help here, but in practice, it is not as simple as might be expected, for interconnect structures to be simulated can be rather complicated due to modern CMOS design rules and multiple materials involved. Also only limited information about material properties required for predictive EM simulation is provided by the foundry.

A measurement-based design approach to avoid the difficulty associated with layout parasitics is shown in Figure 3.2. In this approach, every circuit component is designed to have standard transmission-line (TL) interfaces and is characterized at those interfaces. The standard TL has sidewalls, and unwanted interference between neighboring components is minimal. Thus, all pre-characterized devices can be connected with each other at the interfaces by just bonding as shown in Figure 3.2. No post-LPE is necessary. We called this type of design "bond-based design" [2,3].

3.1.2 Power-line decoupling

Power-supply line decoupling for wideband mmw circuits is a major challenge since commonly used decoupling capacitors exhibit self-resonance. Decoupling capacitors

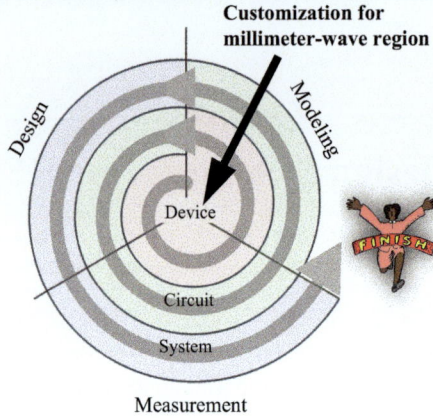

Figure 3.1   *Millimeter-wave CMOS chip development process. Copyright 2015 IEICE. Reproduced, with permission, from [1]*

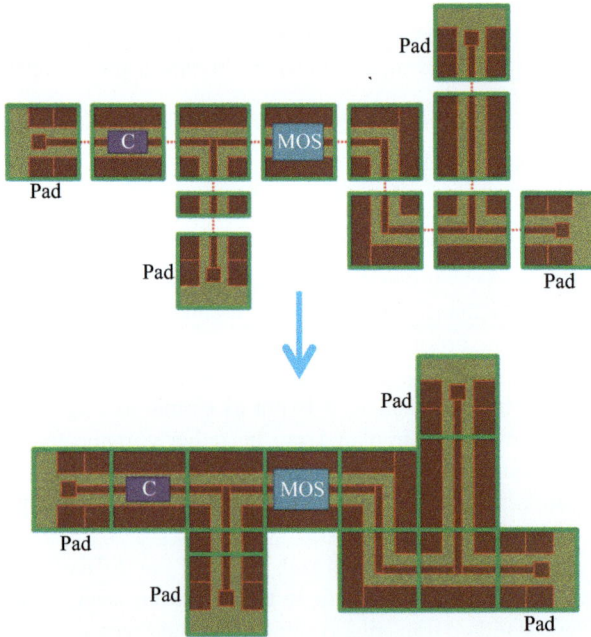

Figure 3.2   *Concept of bond-based design. Pre-characterized component cells having standard transmission-line interfaces are bonded together. Copyright 2015 IEICE. Reproduced, with permission, from [1]*

become inductive above the self-resonant frequency, and power-supply rail shunting (Figure 3.3) does not work reliably at mmw frequencies. To solve this problem, TL with very low characteristic impedance can be utilized [4,5], as shown in Figure 3.4. It serves as a wideband decoupling device as well as power line. Why can a TL with

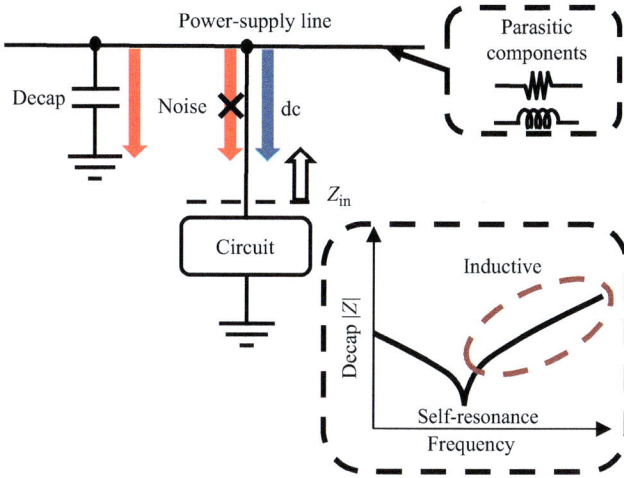

Figure 3.3   Power-supply line shunting with a decoupling capacitor (decap). The impedance of the decap exhibits self-resonance. Copyright 2015 IEICE. Reproduced, with permission, from [1]

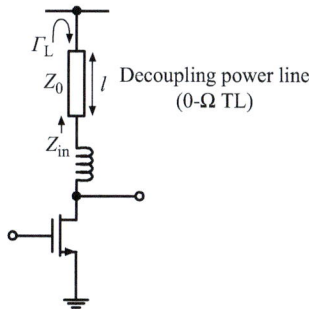

Figure 3.4   0-Ω TL ac-decouples the circuit from the usually unpredictable and uncontrollable far-end termination Γ_L at the power-supply line. Its near-end input impedance Z_{in} approaches the characteristic impedance Z_0 for large l, regardless of the value of Γ_L. Ideally, $Z_0 \to 0\,\Omega$. Copyright 2015 IEICE. Reproduced, with permission, from [1]

low characteristic impedance be a decoupling device? Generally, the input impedance of a terminated *lossy* TL on a Smith chart spirals into its characteristic impedance Z_0, regardless of the far-end termination when its length or frequency is swept [6], as shown in Figure 3.5. The lossier (larger attenuation constant), the faster the input impedance falls onto the characteristic impedance. Obviously, if $Z_0 \simeq 0\,\Omega$, the input impedance also approaches $0\,\Omega$, which is the desired behavior. We call this type of TL design to have as low a characteristic impedance as possible as "the zero-Ohm TL (0-Ω TL)" [7]. The characteristic impedance should be less than $1\,\Omega$ for best decoupling performance [4].

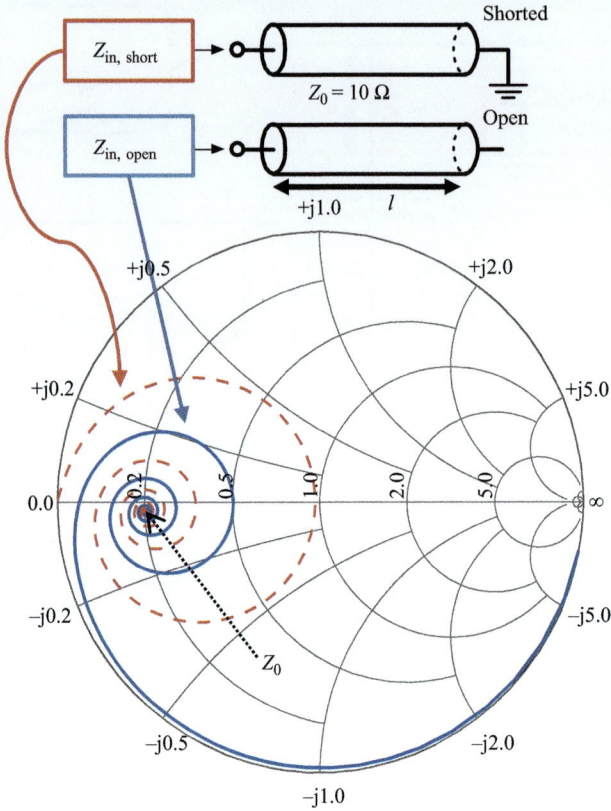

Figure 3.5 *Input impedance Z_{in} of a lossy transmission line converges at its characteristic impedance Z_0 as its length l or the frequency f is swept, regardless of how the far end of the line is terminated. In this example, $Z_0 \simeq 10\,\Omega$, $5\,\mu m \leq l \leq 5\,mm$, and $f = 60\,GHz$. Copyright 2015 IEICE. Reproduced, with permission, from [1]*

How can we achieve low characteristic impedance? Characteristic impedance Z_0 at an angular frequency ω is given by

$$Z_0 = \sqrt{\frac{R + j\omega L}{G + j\omega C}}, \tag{3.1}$$

where R, L, G, and C are the per-unit-length RLGC parameters [6] (Figure 3.6). Here, the propagation constant γ is given by

$$\gamma = \sqrt{(R + j\omega L)(G + j\omega C)}, \tag{3.2}$$

and $\alpha = \Re(\gamma)$ and $\beta = \Im(\gamma)$ are the attenuation and phase constants, respectively. Since the combination of low dc resistance (for low-loss dc power supply) and high

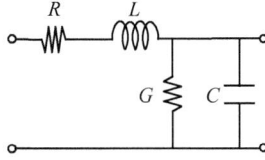

Figure 3.6   RLGC model of a unit-length transmission line section. Copyright 2015 IEICE. Reproduced, with permission, from [1]

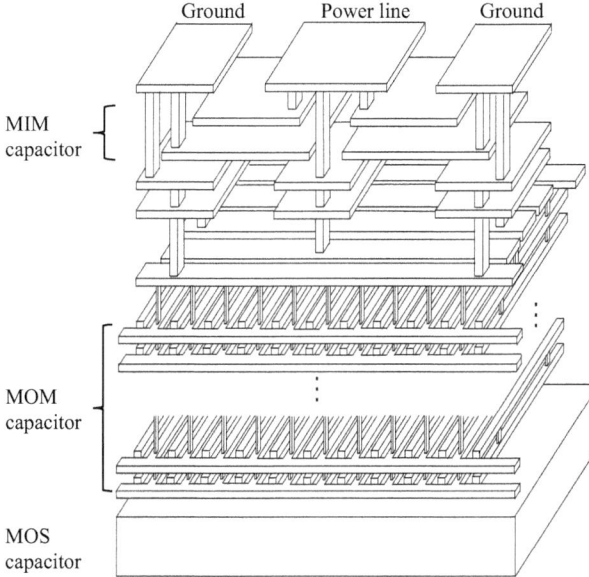

Figure 3.7   Structure of the 0-Ω TL. Copyright 2015 IEICE. Reproduced, with permission, from [1]

RF attenuation (large α) makes a good decoupling device [8,9], the per-unit-length resistance R should be as small as possible for low dc loss, and capacitance C should be as large as possible for small $|Z_0|$ and large α. In CMOS process, 0-Ω TL is typically built as shown in Figure 3.7. Here, thick metal layers are used for the main dc path for small R. Three types of capacitors are used to give a large C: MIM (metal–insulator–metal) parallel-plate capacitors, MOM (metal–oxide–metal) interdigital capacitors and MOS (metal–oxide–semiconductor) capacitors. Since the 0-Ω TL is difficult to characterize due to its low Z_0, we characterized it by measuring its input impedance Z_{in} using specially designed test structures [8]. Measured input impedances of open and shorted stubs of various lengths are shown in Figure 3.8. Z_{in} stays below 2 Ω, which is nearly short, up to 325 GHz. The longer the stub, the faster Z_{in} falls onto Z_0. Since Z_0 of power distribution TL can be used as a performance figure-of-merit [4], Table 3.1 shows comparison of $|Z_0|$ of CMOS-decoupling power lines.

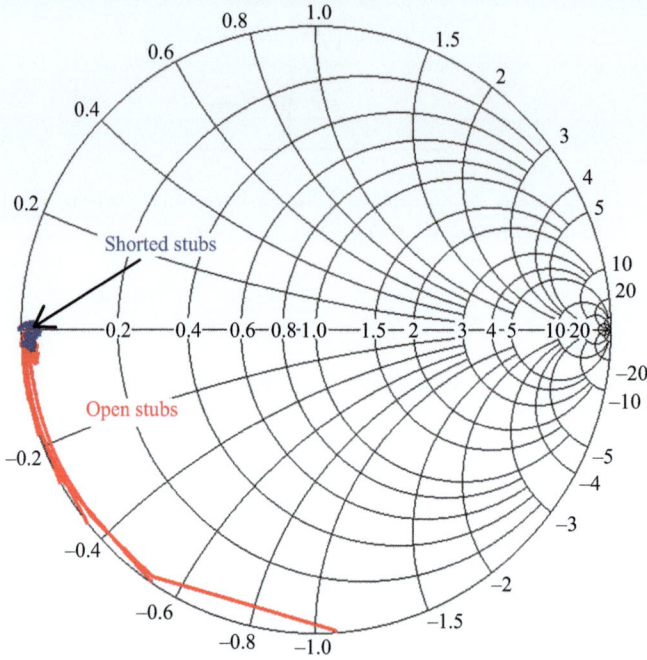

Figure 3.8   Input impedance Z_{in} of open (red) and shorted (blue) stubs (Figure 3.5) of lengths 33, 66, 99, 132, 198, 297, and 396 µm at 1–170 GHz and 220–325 GHz. Copyright 2015 IEICE. Reproduced, with permission, from [1]

Table 3.1   Characteristic impedance Z_0 of CMOS decoupling power lines

	$|Z_0|$ @60 GHz	$|Z_0|$ @300 GHz
[5]	2.2 Ω	–
[7]	1.5 Ω	–
[10]	1.6 Ω	–
[11]	"around 1 Ω"	–
Our work [9]	0.7 Ω	1.6 Ω

3.2   Measurement and de-embedding

3.2.1   *Probing in on-wafer measurement [12]*

Precise measurement is also important for robust device modeling. One of the techniques of precise measurement is position control of the high-frequency probe. Since the wavelength decreases with increasing frequency, even a small fluctuation of the probing position on a pad can affect measurement precision. We use scotch-tape markers on the display, monitoring the silicon chip surface with a microscope, as shown in Figure 3.9. By this technique, the probing position and scratch shapes look almost the same even if two probes are moved manually for different distances between probes, as shown in Figure 3.10. This is because both the start and end points of the probe skating are well controlled approximately within one micrometer. Note that the lateral resolution of landing the probes shown in the *y*-direction is also controlled using chip positioning markers on the display.

3.2.2 De-embedding

3.2.2.1 Basic idea of de-embedding

De-embedding is one of the major factors affecting measurement accuracy. Here we would like to review the outline of de-embedding and the TRL widely method used at ultrahigh frequencies.

Figure 3.11 shows an overview of de-embedding. Basically, a device under test (DUT) is measured on two ports using a vector network analyzer (VNA). Since the

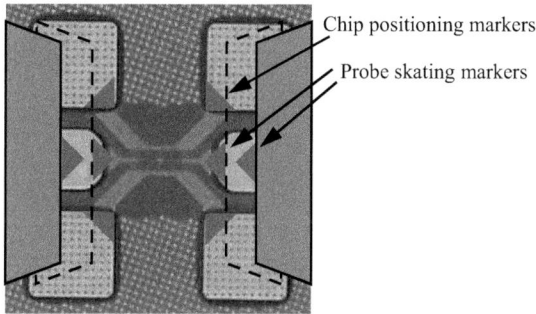

Figure 3.9   *Scotch-tape markers on the display to control the positions of probes. Copyright 2013 IEEE. Reproduced, with permission, from [12]*

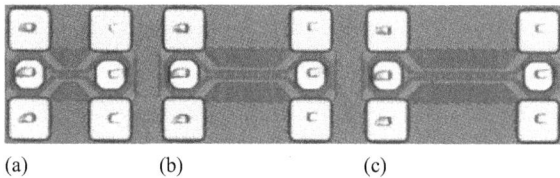

(a)   (b)   (c)

Figure 3.10   *Chip micrograph after probing of on-chip devices: (a) thru dummy, (b) transmission line of 80 μm length, and (c) transmission line of 120 μm length. Positions as well as shapes of scratches made by probing are almost the same on pads even though the distances of the two probes are different among (a) to (c). Positions of probes and pads are aligned by scotch-tape markers attached on a display monitoring a chip surface. Copyright 2013 IEEE. Reproduced, with permission, from [12]*

measurement pad is attached to the DUT, the measurement result also includes the characteristics of the pad. For two ports, there are two pads, each with input and output. Therefore, in general, a four-port embedding network including mutual interference of two pads is formed. This can be represented by a 4×4 matrix. If the 4×4 matrix can be known by some method, the characteristics of the DUT can be obtained by calculation. However, it is difficult to obtain this 4×4 matrix directly by measurement. Therefore, the embedding network is approximately obtained.

The most fundamental method is to assume an equivalent circuit in the embedding network as shown in Figure 3.12. Find the equivalent circuit parameters and remove

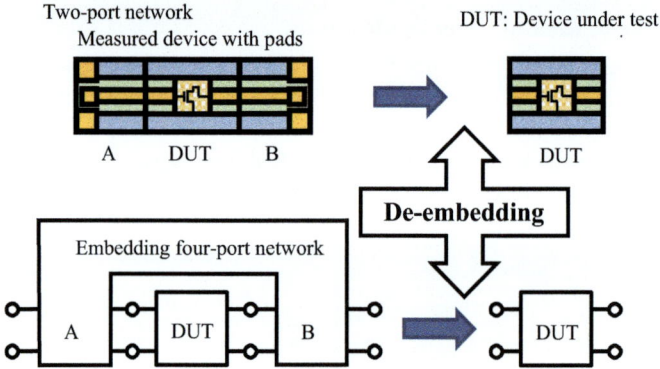

Figure 3.11   *De-embedding is to remove the embedding four-port network for obtaining the DUT characteristics*

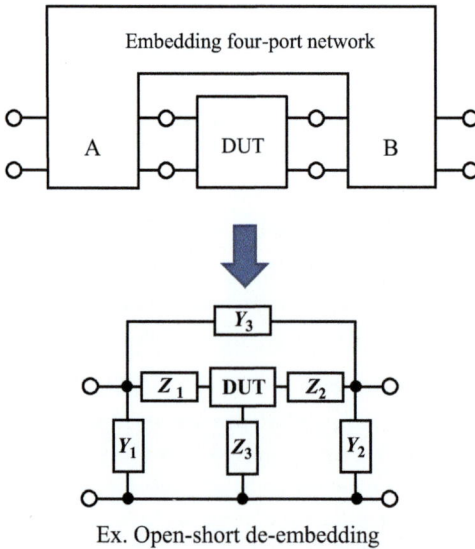

Ex. Open-short de-embedding

Figure 3.12   *De-embedding based on equivalent circuits*

the embedding network from the measurement results. Thus, the characteristics of the DUT are obtained. In Figure 3.12, an open short method is shown. There are two problems to apply this method to high frequencies. One is that ideal open and short cannot be made on the substrate. Since finite admittance and conductance appear, the error increases at high frequencies. In addition, since the characteristic becomes distributed components when the frequency becomes high, the error becomes large in the equivalent circuit based on the lumped components. Therefore, in the mmw band, de-embedding by this equivalent circuit is not used much.

In place of the equivalent circuit method, the cascade matrix method shown in Figure 3.13 is used. This assumes that the influence between the left and right pads can be ignored. If we ignore the influence of each other's pads, the non-diagonal 2 × 2 sub-matrices in the 4 × 4 matrix of the embedding network become zero matrices, and the left and right pads can be treated as independent 2 × 2 matrices. For ultrahigh frequencies, TLs such as microstrip lines are used, and the ground plane is placed near the signal plane. This ground makes mutual interference small enough to be negligible. Therefore, the cascade method is widely used. In the case of the cascade method, a cascade matrix T is used. T is obtained from the S matrix as follows:

$$T = \frac{1}{s_{21}} \begin{bmatrix} -\Delta & s_{11} \\ -s_{22} & 1 \end{bmatrix}, \tag{3.3}$$

where $\Delta = s_{11}s_{22} - s_{12}s_{21}$. If each matrix is defined as shown in Figure 3.14,

$$T_{\text{MEAS}} = T_{\text{A}}T_{\text{DUT}}T_{\text{B}} \tag{3.4}$$

is obtained, and de-embedding can be performed by

$$T_{\text{DUT}} = T_{\text{A}}^{-1}T_{\text{MEAS}}T_{\text{B}}^{-1}. \tag{3.5}$$

The DUT characteristics are obtained by removing embedding cascade matrices.

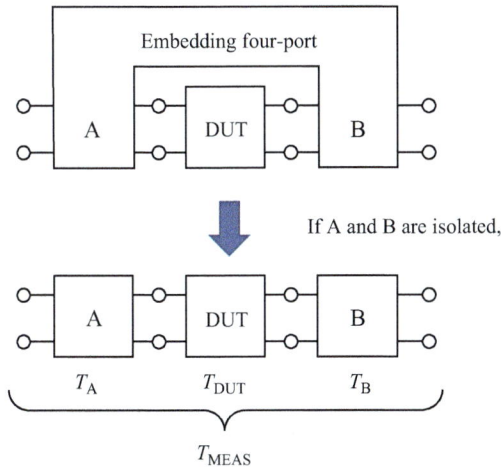

Figure 3.13   *De-embedding based on cascade matrix*

3.2.2.2 Thru–reflect–line (TRL)

The most popular de-embedding of the cascade method is TRL. TRL was proposed in 1979 by Engen and Hoer. In TRL, thru, reflect, and line are used as shown in Figure 3.14. If reflect has no transmission and is symmetrical, impedance need not be known.

Let us begin the calculation in TRL. As shown in Figure 3.15, in TRL, the propagation constant is first obtained using thru and line. Here, T_D is defined as

$$T_D \equiv T_L T_T^{-1}. \tag{3.6}$$

From $T_L = T_A T_{TL} T_B$, obtain $T_{TL} = T_A^{-1} T_D T_A$. If the reference impedance of the T matrix is equal to the characteristic impedance of the TL,

$$T_D \equiv T_L T_T^{-1}. \tag{3.7}$$

$$T_{TL} = \begin{bmatrix} e^{-\gamma l} & 0 \\ 0 & e^{\gamma l} \end{bmatrix} \tag{3.8}$$

becomes the diagonal matrix because there is no reflection. Thus, $e^{-\gamma l}$ is derived from the eigenvalue λ of T_D, when scattering matrix transformed from T_D is symmetric. Finally, we obtain

$$\gamma = \frac{-\ln(\lambda) + j2\pi k}{l} \tag{3.9}$$

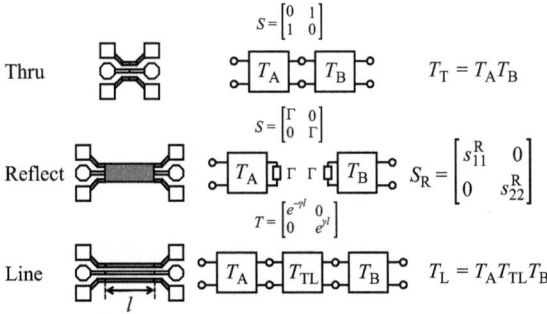

Figure 3.14 TRL (thru–reflect–line)

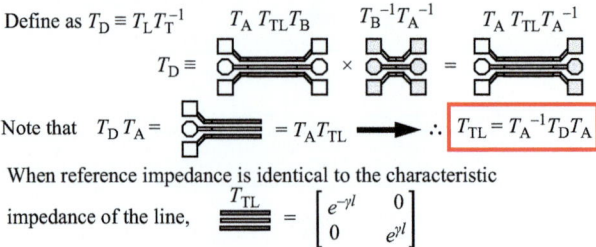

Figure 3.15 Find propagation constant γ of TL

where k is an integer giving correct delay. Here, let

$$T_A = r_1 \begin{bmatrix} a_1 & b_1 \\ c_1 & 1 \end{bmatrix}, T_B = r_2 \begin{bmatrix} a_2 & b_2 \\ c_2 & 1 \end{bmatrix} \text{ and } T_T = r_t \begin{bmatrix} a_t & b_t \\ c_t & 1 \end{bmatrix}. \tag{3.10}$$

Thus, we obtain

$$c_2 = \frac{c_t - a_t c_1 / a_1}{1 - b_t c_1 / a_1}, \tag{3.11}$$

$$\frac{b_2}{a_2} = \frac{b_t - b_1}{a_t - b_1 c_t} \quad \text{and} \tag{3.12}$$

$$a_1 a_2 = \frac{a_t - b_1 c_t}{1 - b_t c_a / a_1}. \tag{3.13}$$

From reflect,

$$s_{11}^R = \frac{a_1 \Gamma + b_1}{c_1 \Gamma + 1} \tag{3.14}$$

a_1 is calculated as

$$a_1 = \frac{s_{11}^R - b_1}{\Gamma \left(1 - (s_{11}^R c_1 / a_1)\right)}. \tag{3.15}$$

Similarly, we obtain

$$a_2 = \frac{s_{22}^R - b_2}{\Gamma \left(1 - (s_{22}^R b_2 / a_2)\right)}. \tag{3.16}$$

Since

$$\frac{a_1}{a_2} = \frac{\left(s_{11}^R - b_1\right) \left(1 + (s_{22}^R b_2 / a_2)\right)}{\left(s_{22}^R + c_2\right) \left(1 + (s_{11}^R c_1 / a_1)\right)}, \tag{3.17}$$

we obtain

$$a_1 = \pm \sqrt{\frac{(s_{11}^R - b_1)(1 + s_{22}^R b_2 / a_2)(a_t - b_1 c_t)}{(s_{22}^R + c_2)(1 - s_{11}^R c_1 / a_1)(1 - b_t c_1 / a_1)}}, \quad \text{and}$$

$$a_2 = \frac{(a_t - b_1 c_t)}{a_1 (1 - b_t c_1 / a_1)}. \tag{3.18}$$

Sign of a_1 is determined by the consistency with (3.15). Now,

$$\begin{bmatrix} a_1 & b_1 \\ c_1 & 1 \end{bmatrix} = \frac{T_A}{r_1} \quad \text{and} \quad \begin{bmatrix} a_2 & b_2 \\ c_2 & 1 \end{bmatrix} = \frac{T_B}{r_2} \tag{3.19}$$

are determined. However, r_1 and r_2 are not individually determined although its product $r_1 r_2 (= r_t)$ is obtained from thru. From T_A / r_1, T_B / r_2, and $r_1 r_2$, DUT is de-embedded as follows:

$$T_{DUT} = T_A^{-1} T_{MEAS} T_B^{-1} = \frac{1}{r_1 r_2} \left(\frac{1}{r_1} T_A\right)^{-1} T_{MEAS} \left(\frac{1}{r_2} T_B\right)^{-1} \tag{3.20}$$

3.2.2.3   Multi-line TRL in terahertz [1]

At above 100 GHz, TRL (thru-reflect-line) de-embedding is commonly used [13]. In particular, multiline TRL [14] gives reasonable de-embedding results since weighted averaging over multiple lines is performed.

The Z_0-referenced S matrix of a length, ΔL, of TL is given by

$$\mathbf{S}_{(Z_0)} = \begin{bmatrix} 0 & e^{-\gamma \Delta L} \\ e^{-\gamma \Delta L} & 0 \end{bmatrix} \tag{3.21}$$

$$= \begin{bmatrix} 0 & e^{-\alpha \Delta L} e^{-j\beta \Delta L} \\ e^{-\alpha \Delta L} e^{-j\beta \Delta L} & 0 \end{bmatrix}. \tag{3.22}$$

The loci of transmission characteristics S_{21} of moderately lossy lines will look as shown in Figure 3.16 on a polar chart. The attenuation constant α can be extracted from the magnitude $|S_{21}|$ and the phase constant β from the argument $\arg S_{21}$. In the context of applying TRL, ΔL corresponds to the line-length difference between a pair of lines.

Actual TLs (except those for decoupling) are designed to be as low-loss as possible. As a result, $|S_{21}| = e^{-\alpha \Delta L}$ [see (3.22)] stays close to unity for typical values of $\Delta L \lesssim 1$ mm. In such a case, precise estimation of α is difficult especially at high frequencies where all sorts of "noise" present in measured S parameters can be large and nearly comparable to $1 - |S_{21}|$. Figure 3.17 shows measured α's extracted by multiline TRL from different line sets. Not only are the resultant α's highly dispersed especially at frequencies above 140 GHz, also the frequency dependence is very erratic. In fact, similar erratic behavior of α at mmw frequencies can often be found in published papers [15–17]. Such erratic behavior cannot easily be reproduced by EM simulation. Aside from measurement equipment noise, spurious modes [18] and crosstalk [20] might be contributing to the "noise" in measured S parameters. To suppress the noise and extract the attenuation constant associated with the lowest

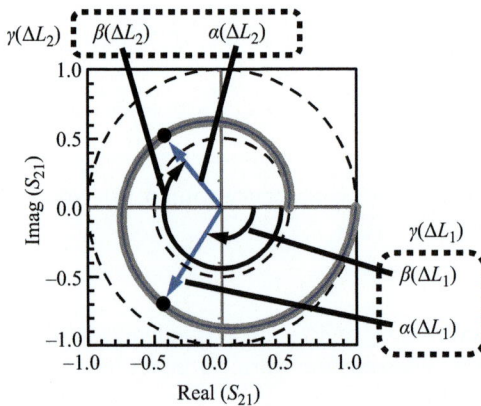

Figure 3.16   *Frequency-dependence of S_{21} of moderately lossy transmission line of lengths ΔL_1 and ΔL_2. $\alpha \Delta L = -\ln |S_{21}|$ and $\beta \Delta L = -\arg S_{21}$. Copyright 2015 IEICE. Reproduced, with permission, from [1]*

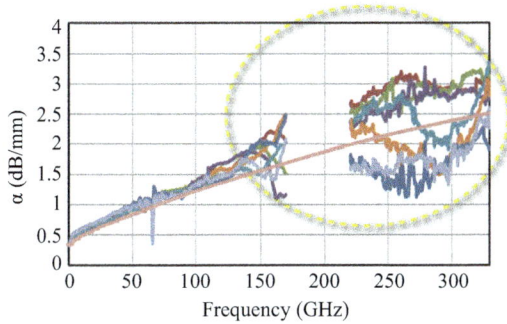

Figure 3.17   Attenuation constant α of a TL extracted by multiline TRL from different line sets. The largest and the smallest values of ΔL were 1.2 and 0.6 mm, respectively. Copyright 2015 IEICE. Reproduced, with permission, from [1]

Table 3.2   Lines sets for multiline TRL

Line set	I	II	III	IV
Line length				
$0\,\mu m$	o	o	o	o
$400\,\mu m$	o	o	o	o
$700\,\mu m$	o	o	o	o
$2,000\,\mu m$	o	o	o	
$3,800\,\mu m$	o	o		
$8,000\,\mu m$	o			
$\Delta L_{max}\,[\mu m]$	8,000	3,800	2,000	700

order propagation mode, the longest line should be very long, leading to a much larger ΔL value than a typical value of about 1 mm. We tried multiline TRL with the line sets shown in Table 3.2, the longest line being 8 mm. The chip micrograph is shown in Figure 3.18. Estimated α and β are shown in Figure 3.19 [21]. The larger the ΔL_{max}, the more well-behaved α is. When $\Delta L_{max} = 8$ mm, the α is very well-behaved. This result suggests that precise de-embedding and reference-plane shifting are possible only if very long lines are used in multiline TRL.

3.2.2.4   Split I for low frequency [19]

Introduction:
A TL is characterized by its propagation constant γ and characteristic impedance Z_0. The γ can be calculated directly from measured S parameters. In contrast, whether the Z_0 can be calculated from the S parameters depends on the knowledge about the reference impedance Z_{ref} of measurement planes.

In the case of an on-chip TL, the TL is embedded in an embedding network consisting typically of probe pads and access lines. The reference planes must be brought

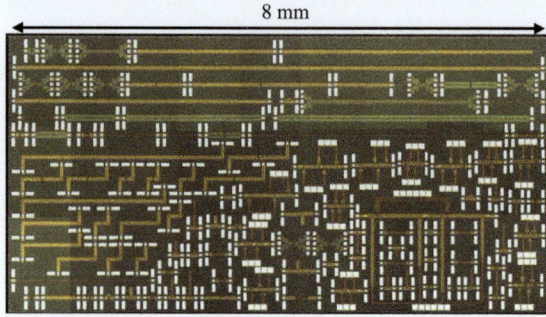

Figure 3.18   Micrograph of a 65-nm CMOS transmission line chip. Copyright 2015 IEICE. Reproduced, with permission, from [1]

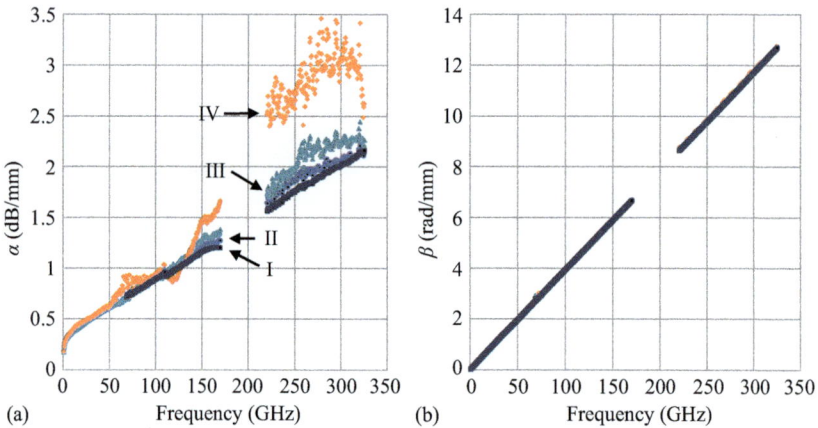

Figure 3.19   Characteristics of the shielded microstrip line extracted from different line sets (Table 3.2) by multiline TRL: (a) attenuation constant α and (b) phase constant β. All four results of β are indistinguishable on this scale. Copyright 2015 IEICE. Reproduced, with permission, from [1]

onto the TL either by a first-tier VNA calibration using on-chip calibration standards or by a second-tier de-embedding procedure. The latter typically follows first-tier probe-tip calibration using a commercially available impedance-standard substrate. If the subsequent de-embedding of the embedded TL is done using some equivalent-circuit-based technique, the reference resistance R_{ref} (equal to 50 Ω in most cases) set by the first-tier calibration is transferred to the new reference planes. In such a case, Z_0 can be calculated directly from the de-embedded TL's S parameters. On the other hand, if the reference planes are defined by the first- or second-tier TRL (thru-reflect-line) [14], the Z_{ref} equals Z_0, which is complex and unknown. In that case, no information about Z_0 is obtained from S-parameter measurement alone. A standard approach to finding the value of Z_0 involves low-frequency capacitance or resistance measurement using different measurement samples than those for S-parameter measurements [22]. The Z_0 is then estimated from (3.23) using the information thus obtained and the γ that resulted from TRL [23].

$$Z_0 = \frac{\gamma}{G + j\omega C} \simeq \frac{\gamma}{j\omega C}, \tag{3.23}$$

where $\gamma = \sqrt{(R + j\omega L)(G + j\omega C)}$ and R, L, G, and C are the per-unit-length RLGC parameters of the TL. However, due to its extra measurement complexity, S-parameter-only determination of Z_0 is often preferred to this approach.

In this section, we examine three equivalent-circuit-based techniques for de-embedding TLs, performed following probe-tip calibration, assuming that the calibration is reliable. Based on the analysis, we suggest some possible alternative techniques.

TL de-embedding techniques:
We assume here that a length, ℓ, of TL to be de-embedded sits in between two embedding two ports as shown in Figure 3.20(a), each of them consisting of pads and access lines. Another device measured for the purpose of de-embedding has no embedded TL, as shown in Figure 3.20(b). We will refer to these samples as LINE and THRU, respectively. We assume that our LINE and THRU have left–right symmetry, and hence the symbols F and ꟻ for representing the left and the right halves of THRU. We also assume that the LINE and the THRU have symmetric Y and Z matrices (reciprocity assumption). This should be valid at not-so-high frequencies where probe-tip calibration is usable. We further assume that there is no leakage path between the two measurement ports that bypass the networks shown in Figure 3.20 (no-crosstalk assumption). LINE and THRU can be expressed in terms of ABCD (or F) matrices as follows;

$$\mathbf{F}_L = \mathbf{F}_F \mathbf{F}_{TL} \mathbf{F}_ꟻ, \quad \text{[Figure 3.20(a)]} \tag{3.24}$$

$$\mathbf{F}_T = \mathbf{F}_F \mathbf{F}_ꟻ. \quad \text{[Figure 3.20(b)]}. \tag{3.25}$$

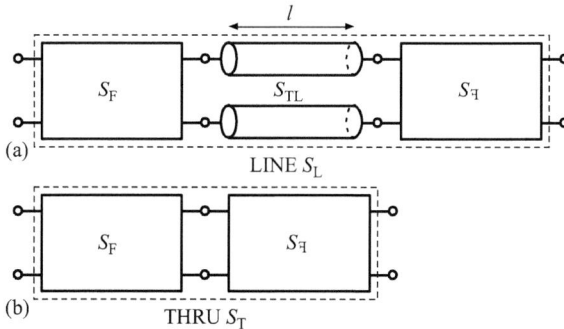

Figure 3.20   (a) Symmetric LINE and (b) symmetric THRU. Copyright 2015 IEEE. Reproduced, with permission, from [19].

If F_F can somehow be determined from measurement data, the TL can be de-embedded as follows:

$$F_{TL} = F_F^{-1} F_L F_\daleth^{-1}. \tag{3.26}$$

De-embedding with split-Π thru: A popular technique of determining Y_F (Y matrix of the left half of THRU) is to represent the THRU by the split-Π equivalent, shown in Figure 3.21(a). See [24] and references therein. Then,

$$Y_{F,\Pi} = \begin{bmatrix} Y_{T11} - Y_{T21} & 2Y_{T21} \\ 2Y_{T21} & -2Y_{T21} \end{bmatrix}, \tag{3.27}$$

where Y_{Tij} are the elements of Y_T (Y matrix of THRU), which can be measured. The TL can then be de-embedded by applying (3.26), leading to $F_{TL,\Pi}$.

De-embedding with split-T thru: Another technique of determining Z_F (Z matrix of the left half of THRU) is to represent the THRU by the split-T equivalent, shown in Figure 3.21(b). Then,

$$Z_{F,T} = \begin{bmatrix} Z_{T11} + Z_{T21} & 2Z_{T21} \\ 2Z_{T21} & 2Z_{T21} \end{bmatrix}, \tag{3.28}$$

where Z_{Tij} are the elements of Z_T (measured Z matrix of THRU). Again, the TL can be de-embedded by applying (3.26), leading to $F_{TL,T}$. As was pointed out in [24], $F_{TL,T} \neq F_{TL,\Pi}$ in general despite $F_T = F_{F,T} F_{\daleth,T} = F_{F,\Pi} F_{\daleth,\Pi}$.

De-embedding by admittance cancellation by swapping: Yet another technique for de-embedding TLs, proposed in [25], makes the "lumped-pad assumption," in which

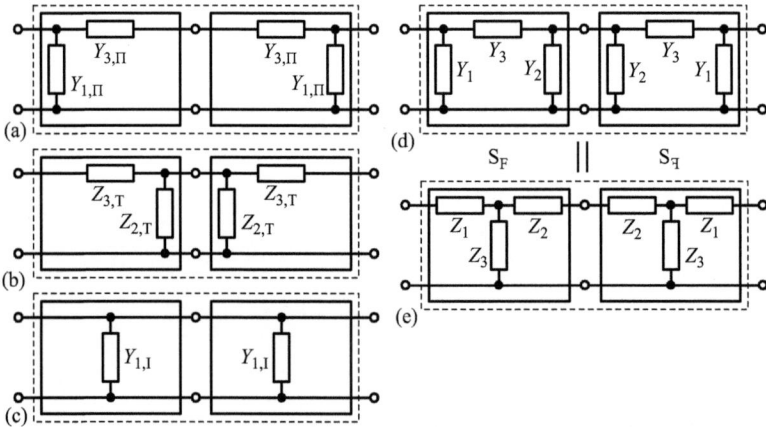

Figure 3.21   *Possible representations of a symmetric* THRU, *used for de-embedding: (a) split* Π, *(b) split T, (c) split I, (d) double* Π, *and (e) double T. Only left (right) halves of (d) and (e) (but not those of (a)–(c)) can correctly represent* S_F ($S_\daleth$), *the true left (right) half of the* THRU. *Copyright 2015 IEEE. Reproduced, with permission, from [19]*

each half of the THRU is assumed to be representable by a single shunt branch as shown in Figure 3.21(c). However, the actual de-embedding procedure of [25] is not quite the same as somehow determining the value of $Y_{I,1}$ in Figure 3.21(c) and applying (3.26). Instead, [25] first introduces the length-difference ABCD matrix

$$\mathsf{F}_H \triangleq \mathsf{F}_L \mathsf{F}_T^{-1} = \mathsf{F}_F \mathsf{F}_{TL} \mathsf{F}_F^{-1}, \tag{3.29}$$

and, using its Y matrix Y_H, de-embeds the TL by

$$\mathsf{Y}_{TL,ACS} = \left(\mathsf{Y}_H + \overline{\mathsf{Y}_H}\right)/2, \tag{3.30}$$

where the overline denotes the port swapping ($1 \leftrightarrow 2$) or left–right mirroring operation. If the lumped-pad assumption is valid, the shunt admittance $Y_{I,1}$ is canceled out by (3.30). But the lumped-pad assumption is usually hard to justify. In general, $\mathsf{Y}_{TL,ACS} \neq \mathsf{Y}_{TL,\Pi}$ and $\mathsf{Y}_{TL,ACS} \neq \mathsf{Y}_{TL,T}$ [24].

Comparative analysis:
In the numerical examples that follow, we use synthesized data that are meant to be reasonably realistic models of a TL ($Z_0 \simeq 55\,\Omega$) and ground-signal-ground (GSG) pads. The simulation models used to generate the data are similar to the ones used in [24]. This is to highlight the essential features of the de-embedding formulations when none of our assumptions (validity of probe-tip calibration, symmetry, reciprocity, and no-crosstalk) is violated and to avoid obfuscation caused by various uncertainties present in real measurement data. The knowledge about the correct de-embedding result also makes comparison of the formulations easier.

The correct value of γ is obtained from the all three techniques [24]. Figure 3.22 shows the real and imaginary parts of the correct and the de-embedded Z_0. Notable features are

1. $\Re(Z_0)$ from "split Π" shows a peculiar downward bend. Similar results can be found in Fig. 12 of [26] and Fig. 3 of [27]. $\Im(Z_0)$ from "split Π" also is far off except at low frequencies.
2. $\Re(Z_0)$ from "split T" shows a peculiar upward bend. Similar results can be found in Fig. 13 of [26] and Fig. 3 of [27]. $\Im(Z_0)$ from "split T" also is far off except at low frequencies.
3. $\Re(Z_0)$ and $\Im(Z_0)$ from "ICS Y" [Equation (3.30)] are well-behaved and show reasonable agreement with the true Z_0 throughout. Such well-behaved results can also be found in [25].

While the split-Π [Figure 3.21(a)] and the split-T [Figure 3.21(b)] work well only at low frequencies where the length of the THRU is very much shorter than the wavelength ($\lesssim \lambda/40$, e.g.), the low-frequency results can be used to extrapolate the Z_0 up to the full measurement frequency range [29]. But clearly, "ICS Y" gives much better results in spite of the lesser number of elements in Figure 3.21(c) that is used to represent a half of the THRU than in Figure 3.21(a) and (b).

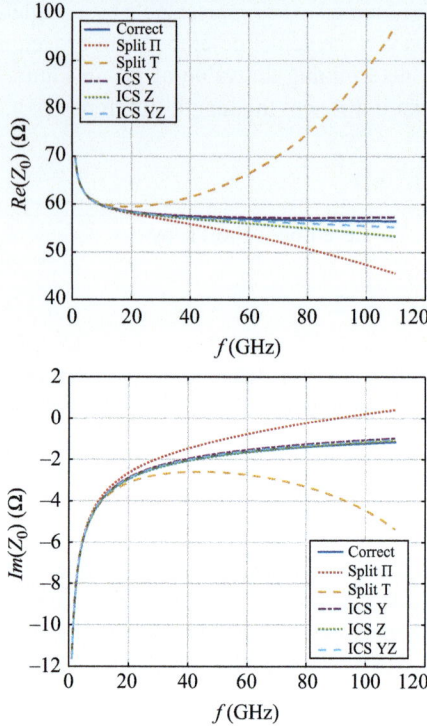

*Figure 3.22   Correct (solid) and estimated (others) $\Re(Z_0)$ and $\Im(Z_0)$. "Split Π"
results from (3.27), "Split T" from (3.28), "ICS Y" from (3.30), "ICS
Z" from (3.35), and "ICS YZ" from (3.36). ICS, Immittance
cancellation by swapping. Copyright 2015 IEEE. Reproduced, with
permission, from [19]*

To see why "ICS Y" outperforms the other two, let us represent the halves of the
THRU as in Figure 3.21(d) and (e) using the reciprocity assumption and the unknowns
Y_i and Z_i ($i = 1, 2, 3$):

$$\mathbf{Y}_F = \begin{bmatrix} Y_1 + Y_3 & -Y_3 \\ -Y_3 & Y_2 + Y_3 \end{bmatrix} \quad \text{[Figure 3.21(d)]} \tag{3.31}$$

$$= \begin{bmatrix} Z_1 + Z_3 & Z_3 \\ Z_3 & Z_2 + Z_3 \end{bmatrix}^{-1} = \mathbf{Z}_F^{-1}. \quad \text{[Figure 3.21(e)]} \tag{3.32}$$

In terms of Y_i and Z_i, (3.27) and (3.28) are

$$\mathbf{Y}_{F,\Pi} = \begin{bmatrix} Y_1 + Y_3 & -Y_3/(1 + Y_2/Y_3) \\ -Y_3/(1 + Y_2/Y_3) & Y_3/(1 + Y_2/Y_3) \end{bmatrix}, \tag{3.33}$$

$$\mathbf{Z}_{F,T} = \begin{bmatrix} Z_1 + Z_3 & Z_3/(1 + Z_2/Z_3) \\ Z_3/(1 + Z_2/Z_3) & Z_3/(1 + Z_2/Z_3) \end{bmatrix}. \tag{3.34}$$

It turns out by using (3.33) and (3.34) that the "de-embedded" TLs actually *embed* the
TL [Figure 3.23(a)] in some rather complicated networks, as shown in Figure 3.23(b)

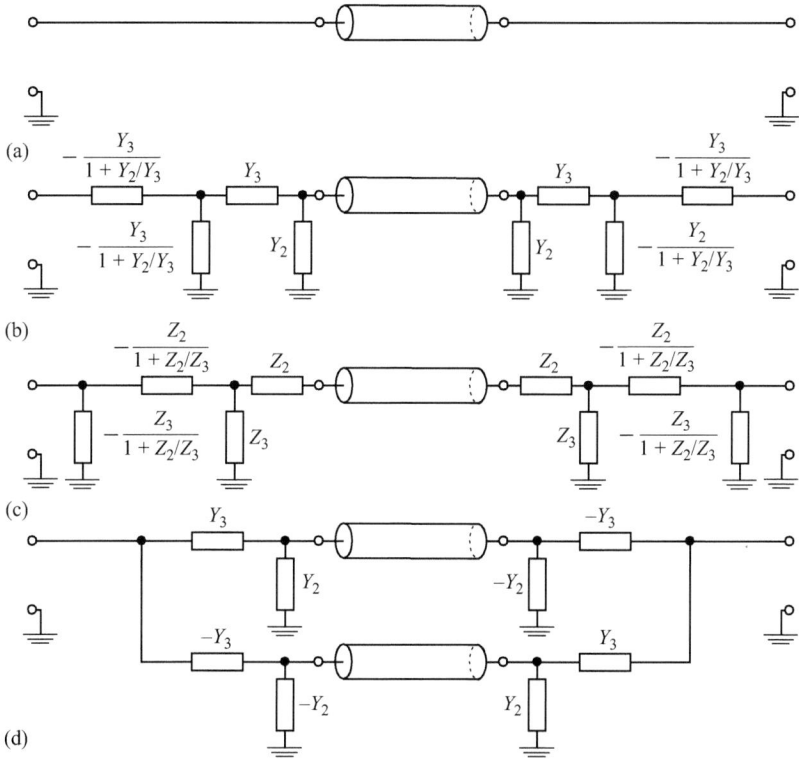

Figure 3.23   *(a) Ideally de-embedded TL, (b) TL de-embedded by split Π [Figure 3.21(a)], (c) TL de-embedded by split T [Figure 3.21(b)], and (d) two de-embedded TLs in parallel, corresponding to the numerator of (3.30). Copyright 2015 IEEE. Reproduced, with permission, from [19]*

and (c). The image impedances of Figure 3.23(b) and (c) are what we saw in Figure 3.22 as the "de-embedded Z_0." The numerator of (3.30) is shown in Figure 3.23(d). While the detailed frequency response of each of these results depends on the actual frequency-dependent values of Y_i and Z_i, which are unknown (not so in the present case), it is not difficult to see that Figure 3.23(d) is closer to the ideal result of Figure 3.23(a) than Figure 3.23(b) and (c).

Given the fact that (3.30) works well, it is possible that its dual expression, (3.35), works well too.

$$Z_{\text{TL,ICS}} = \frac{\left(\mathbf{Z}_{\text{H}} + \overline{\mathbf{Z}}_{\text{H}}\right)}{2}, \tag{3.35}$$

where $\mathbf{Z}_{\text{H}}$ is the Z-matrix of (3.29). Even some form of combination of (3.30) and (3.35) might also work. For example,

$$\mathbf{Y}_{\text{TL,ICS,YZ}} = \frac{\left(\mathbf{Y}_{\text{TL,ACS}} + \mathbf{Z}_{\text{TL,ICS}}^{-1}\right)}{2}. \tag{3.36}$$

The results are also plotted in Figure 3.22.

Conclusion and discussion:

We analyzed three known S-parameter-only (i.e., no additional capacitance measurement) techniques for on-chip TL de-embedding and clarified why the apparently most simplistic one [Equation (3.30)] [25] outperforms the other two [Figure 3.21(a) and (b)]. This seems to be a case of "more does not necessarily mean better." We also suggested possible alternatives formulas, (3.35) and (3.36). Although these do not appear to give significantly better results than (3.30) as far as the example shown in Figure 3.22 is concerned, we confirmed that when the underlying model is different (e.g., the one in [28]), somewhat better result can be obtained from (3.35) or (3.36).

In the actual experimental situation where the correct result is unknown, the choice should be made based on the plausibility of the resulting frequency dependence of $\Re(Z_0)$ and $\Im(Z_0)$. For example, in Figure 3.22, the result from (3.30) actually shows slight upward bend, which might be deemed physically unjustifiable. In such a case, (3.35) or (3.36) would be a reasonable alternative. At high frequencies where the assumptions made herein are violated, Z_0-extrapolation similar to [29] would be necessary. In such a case too (3.30), (3.35), or (3.36) would be useful for building a required model.

3.2.3   Parameter extraction for EM field simulation [1]

To perform accurate EM field analysis at mmw and terahertz frequencies, precise material parameters are required. Although the nominal values of metal conductivities and the real parts of dielectric permittivities are provided by foundries, other parameters that critically affect the performance, including complex dielectric permittivities and effective thicknesses of interconnect layers, are generally not provided. Thus, we developed a systematic method of calibrating process parameters, applicable up to terahertz frequencies [30,31]. Since propagation constants of TLs extracted by multiline TRL using long lines are fairly reliable [21], propagation constants are utilized for extracting process parameters together with the estimates of characteristic impedance [29,32]. A cross-sectional structure of a standard CMOS interconnect layers is shown in Figure 3.24. After applying TRL, process parameters are extracted following the flowchart shown in Figure 3.25. The propagation constants in the entire measurement frequency range and the RLGC parameters (Figure 3.6) at low frequencies are used to determine process parameters. By using four types of microstrip lines shown in Figure 3.26, we determined process parameters. Attenuation constants and

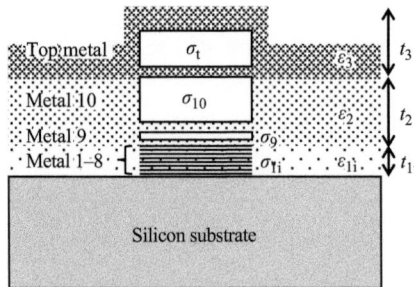

Figure 3.24   Cross-sectional structure of a standard CMOS interconnect layers. Copyright 2015 IEICE. Reproduced, with permission, from [1]

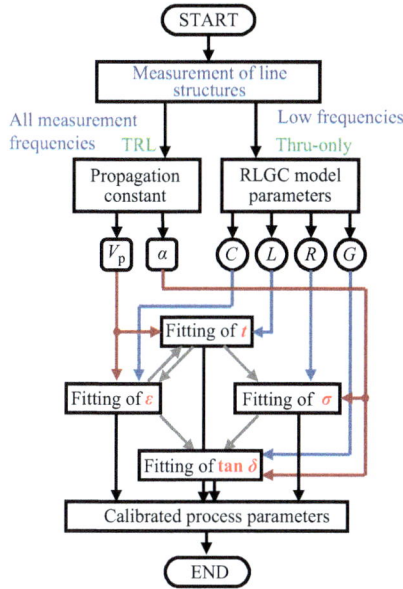

Figure 3.25 *Flowchart of the proposed process parameter calibration method. v_p is the phase velocity. α is the attenuation constant. σ is the metal conductivity. t is the thickness of a dielectric. ε is the real part of the dielectric constant. $\tan \delta$ is the dielectric loss tangent. Copyright 2015 IEICE. Reproduced, with permission, from [1]*

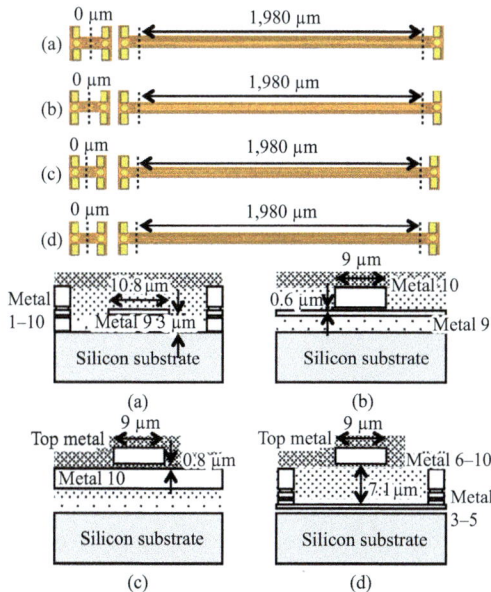

Figure 3.26 *Micrographs of four types of microstrip and their respective cross-sectional structures. Copyright 2015 IEICE. Reproduced, with permission, from [1]*

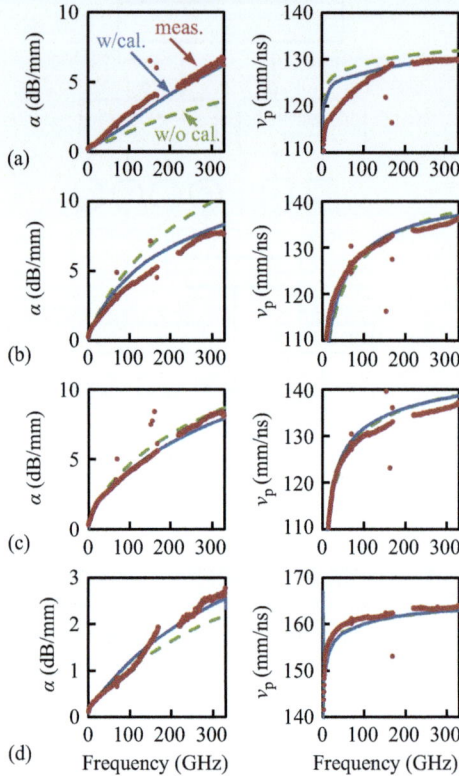

Figure 3.27   *Comparison of the attenuation constant α and the phase velocity*
$v_p = \omega/\beta$ *for the four types of microstrip shown in Figure 3.26,*
obtained from measurement (red) and EM simulation with (blue) and
without (green) process parameter calibration. Copyright 2015
IEICE. Reproduced, with permission, from [1]

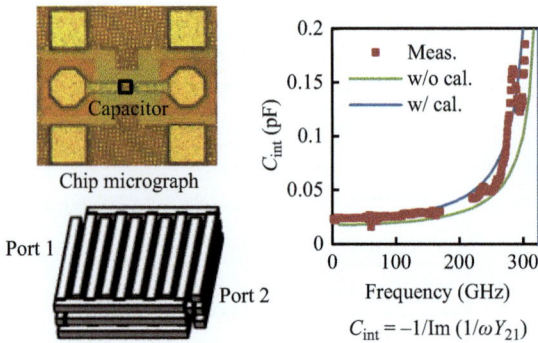

$$C_{int} = -1/Im\,(1/\omega Y_{21})$$

Figure 3.28   *Structure of an interdigital MOM capacitor and the intrinsic*
capacitance obtained from EM simulations and measurement.
Copyright 2015 IEICE. Reproduced, with permission, from [1]

phase velocities obtained from measurement and EM simulation are compared in Figure 3.27 [31]. Measured capacitance of an on-chip interdigital MOM capacitor is compared with that obtained from EM simulation in Figure 3.28. The characteristics calculated with the calibrated process parameters show better agreement with those obtained from the measurement.

3.3 Device modeling

3.3.1 Small-signal equivalent-circuit modeling

In the design of ultrahigh-frequency integrated circuits, devices designed by users, in addition to devices supplied from foundries of integrated circuits, will be used. Since the measurement result of this self-build devices is not suitable for use in the circuit simulation as it is, it is necessary to build a model. In this section, to derive an equivalent circuit of devices systematically, approximation by real rational function using measurement results will be discussed. The result of modeling the small signal characteristics of the mmw device using this method is also demonstrated.

3.3.1.1 Small-signal equivalent circuit of integrated-circuit device

In the design of integrated circuits, design information called a process design kit (PDK) is usually provided from the foundry of the integrated circuit. Circuit simulation is performed based on the provided PDK to confirm the circuit characteristics, and then the design is signed off. However, since PDK is not guaranteed accuracy in the mmw band (30–300 GHz) in many cases, it is necessary to create your own model after measuring the device using a VNA in the mmw band. The reason why modeling is necessary without using measurement results is to ensure convergence, high speed, smooth characteristics (not including measurement noise) in circuit simulation, and characteristics outside the measurement range. Figure 3.29 shows modeling of MIM capacitor as an example, from measurement to de-embedding, S-to-Y parameter conversion, Y parameter to π-type equivalent circuit conversion. These series of procedures are necessary before device modeling, and the accuracy of these procedures greatly affects the result of circuit simulation. Here, since the devices of the integrated circuit are nonlinear, strictly, a nonlinear model is required. In many circuits not limited to amplifiers, however, small signal characteristics often give good approximations, and small signal equivalent circuits are still important. Here we discuss the method of systematically deriving a linear equivalent circuit model from the measurement result.

Model by real rational function:
Figure 3.30 illustrates systematic modeling using real rational functions. If the admittance $Y(\omega)$ of the measurement result is a function of angular frequency ω, the conductance $G(\omega)$ and the susceptance $B(\omega)$ are also the functions of ω. Let it be approximated by a real rational function of a finite order. The order referred to here is the multiplier of angular frequency ω. In this example, the order of the denominator

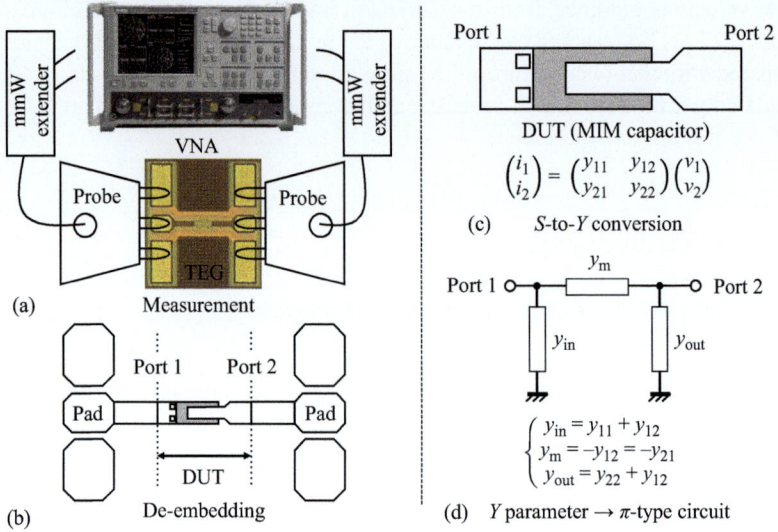

(a)   Measurement

(b)   De-embedding

(c)   S-to-Y conversion

$$\begin{pmatrix} i_1 \\ i_2 \end{pmatrix} = \begin{pmatrix} y_{11} & y_{12} \\ y_{21} & y_{22} \end{pmatrix} \begin{pmatrix} v_1 \\ v_2 \end{pmatrix}$$

(d) Y parameter $\rightarrow$ π-type circuit

$$\begin{cases} y_{\text{in}} = y_{11} + y_{12} \\ y_{\text{m}} = -y_{12} = -y_{21} \\ y_{\text{out}} = y_{22} + y_{12} \end{cases}$$

Figure 3.29 *Flow from measurement to π-type equivalent circuit before equivalent circuit modeling*

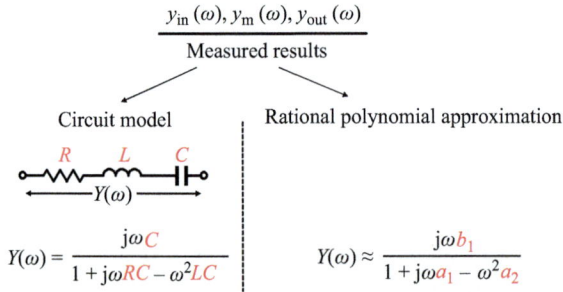

$$Y(\omega) = \frac{j\omega C}{1 + j\omega RC - \omega^2 LC}$$

$$Y(\omega) \approx \frac{j\omega b_1}{1 + j\omega a_1 - \omega^2 a_2}$$

Figure 3.30 *Relationship between equivalent circuit model and real rational function derived from measurement results*

is 2 and the order of the numerator is 1. On the other hand, if the circuit model is a series of R, L, and C, its admittance Y is a function of the angular frequency ω, the denominator order is 2, and the numerator order is 1. Since this is the same as the order of the real rational function, the coefficients b_1, a_1 and a_2 of the real rational function and R, L, C can be mutually converted by comparing the coefficients. That is, if the coefficients of the real rational function are found, the values of the circuit model are determined. If the difference of the order between the denominator and the numerator is 1 or less and the real rational function satisfies positive real condition, there exists a corresponding physical circuit model.

Next, coefficients of the real rational function are derived from the measurement results. The left side of the following equation is the measurement result and the right side is the real rational function.

$$Y(\omega) = G(\omega) + jB(\omega) \approx \frac{j\omega b_1}{1 + j\omega a_1 - \omega^2 a_2}. \tag{3.37}$$

Transfer the right side to the left and arrange the real part and the imaginary part to obtain

$$G(\omega) - \omega B(\omega)a_1 - \omega^2 B(\omega)a_2$$
$$+ j \left\{ B(\omega) + \omega G(\omega)a_1 - \omega^2 B(\omega)a_2 - \omega b_1 \right\} \approx 0. \tag{3.38}$$

To obtain the coefficient by the least squares method, if the sum of the square residuals of all measurement points is set as J,

$$J = \sum_\omega \left| G(\omega) - a_1 \omega B(\omega) - a_2 \omega^2 B(\omega) \right.$$
$$\left. + j \left\{ B(\omega) + a_1 \omega G(\omega) - a_2 \omega^2 B(\omega) - b_1 \omega \right\} \right|^2 \tag{3.39}$$

is obtained. To obtain coefficients a_1, a_2, and b_1 for minimizing this J, (3.39) is partially differentiated with each coefficient.

$$\begin{cases} \dfrac{\partial J}{\partial a_1} = \sum_\omega \left(2\omega^2 \left(G^2(\omega) + B^2(\omega) \right) a_1 - 2\omega^2 G(\omega)b_1 \right) = 0 \\[4mm] \dfrac{\partial J}{\partial a_2} = \sum_\omega \left(2\omega^4 \left(G^2(\omega) + B^2(\omega) \right) a_2 + 2\omega^3 B(\omega)b_1 - 2\omega^2 \left(G^2(\omega) + B^2(\omega) \right) \right) = 0 \\[4mm] \dfrac{\partial J}{\partial b_1} = \sum_\omega \left(-2\omega^2 G(\omega)a_2 + 2\omega^3 B(\omega)a_2 + 2\omega^2 b_1 - 2\omega B(\omega) \right) = 0 \end{cases} \tag{3.40}$$

The matrix form is convenient when obtaining the solution of (3.40) by numerical operation. Therefore, when (3.40) is modified to matrix equation,

$$\sum_\omega \begin{bmatrix} \omega^2 \left(G^2(\omega) + B^2(\omega) \right) & & -\omega^2 G(\omega) \\ & \omega^4 \left(G^2(\omega) + B^2(\omega) \right) & \omega^3 B(\omega) \\ -\omega^2 G(\omega) & \omega^3 B(\omega) & \omega^2 \end{bmatrix} \begin{bmatrix} a_1 \\ a_2 \\ b_1 \end{bmatrix}$$
$$= \sum_\omega \begin{bmatrix} \omega^2 \left(G^2(\omega) + B^2(\omega) \right) \\ -\omega B(\omega) \end{bmatrix} \tag{3.41}$$

is obtained. Solving this gives the coefficients a_1, a_2, and b_1. These coefficients are converted into values of R, L, and C as shown in Figure 3.30. In this way, by using the real rational function, the parameters of the circuit model can be systematically calculated.

Comparison of measured results with modeled results:

Let us compare the results modeled using the above method with the measurement results. Here, the MIM capacitor is modeled by a π-type equivalent circuit as shown in Figure 3.29. The marker in Figure 3.31 shows the frequency characteristic of the mutual admittance y_m obtained from the measurement result of the MIM capacitor. The circuit model consisting of R_m, L_m, and C_m is shown in the upper left box of the figure, and the graph shows its real and imaginary parts together with measurement data. The shunt admittances y_{in} and y_{out} are modeled in the same manner, and the results are shown in Figure 3.32. From these results, the device model can be systematically and accurately derived from the measurement result by using this method.

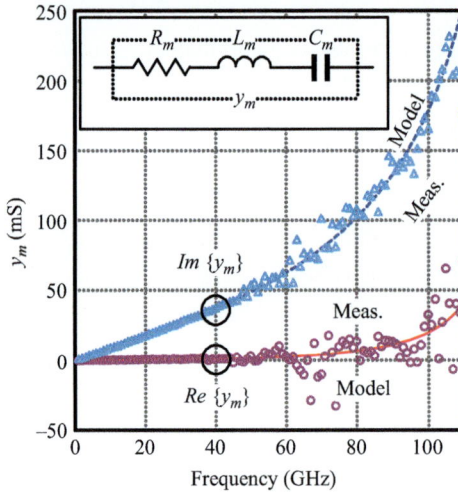

Figure 3.31   *Comparison between model and measurement data (y_m)*

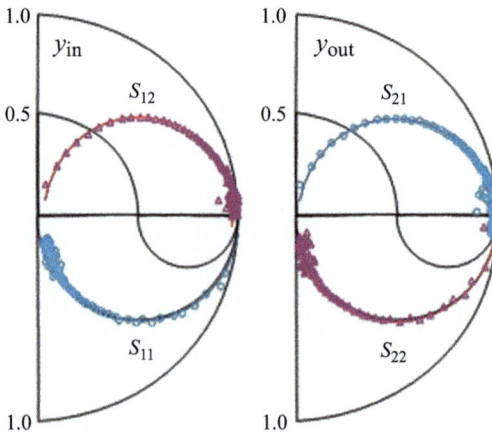

Figure 3.32   *Comparison between results modeled using real rational function and measurement results (y_{in} and y_{out}). Results are shown by S parameters*

Extend order of real rational function:

In the previous chapter, we have discussed the case where the denominator is the second order and the numerator is the first order. However, this method can extend the order of the denominator and the numerator. Note that the difference between the order of the denominator and the order of the numerator needs to be 1 or less if an equivalent circuit consisting of ordinary L, C, and R is assumed.

$$Y(\omega) = \frac{b_0 + j\omega b_1 - \omega^2 b_2 + \cdots}{a_0 + j\omega a_1 - \omega^2 a_2 + \cdots} \tag{3.42}$$

The matrix representation of the equation for calculating coefficients a_i and b_j is given by

$$\begin{bmatrix} AA & BA \\ AB & BB \end{bmatrix} \begin{bmatrix} a_0 \\ a_1 \\ \vdots \\ b_0 \\ b_1 \\ \vdots \end{bmatrix} = 0, \tag{3.43}$$

where

$$AA = \begin{bmatrix} (G^2+B^2) & -(G^2+B^2)\omega^2 & (G^2+B^2)\omega^4 & -(G^2+B^2)\omega^6 & \cdots \\ (G^2+B^2)\omega^2 & -(G^2+B^2)\omega^4 & (G^2+B^2)\omega^6 & \cdots \\ -(G^2+B^2)\omega^2 & (G^2+B^2)\omega^4 & -(G^2+B^2)\omega^6 & (G^2+B^2)\omega^8 & \cdots \\ -(G^2+B^2)\omega^4 & (G^2+B^2)\omega^6 & -(G^2+B^2)\omega^8 & \cdots \\ (G^2+B^2)\omega^4 & -(G^2+B^2)\omega^6 & (G^2+B^2)\omega^8 & -(G^2+B^2)\omega^{10} & \cdots \\ (G^2+B^2)\omega^6 & -(G^2+B^2)\omega^8 & (G^2+B^2)\omega^{10} & \cdots \\ -(G^2+B^2)\omega^6 & (G^2+B^2)\omega^8 & -(G^2+B^2)\omega^{10} & (G^2+B^2)\omega^{12} & \cdots \\ \vdots & \vdots & \vdots & \ddots \end{bmatrix}$$

$$AB = \begin{bmatrix} -G & B\omega & G\omega^2 & -B\omega^3 & -G\omega^4 & B\omega^5 & G\omega^6 & \cdots \\ -B\omega & -G\omega^2 & B\omega^3 & G\omega^4 & -B\omega^5 & -G\omega^6 & B\omega^7 & \cdots \\ G\omega^2 & -B\omega^3 & -G\omega^4 & B\omega^5 & G\omega^6 & -B\omega^7 & -G\omega^8 & \cdots \\ B\omega^3 & G\omega^4 & -B\omega^5 & -G\omega^6 & B\omega^7 & G\omega^8 & -B\omega^9 & \cdots \\ -G\omega^4 & B\omega^5 & G\omega^6 & -B\omega^7 & -G\omega^8 & B\omega^9 & G\omega^{10} & \cdots \\ -B\omega^5 & -G\omega^6 & B\omega^7 & G\omega^8 & -B\omega^9 & -G\omega^{10} & B\omega^{11} & \cdots \\ G\omega^6 & -B\omega^7 & -G\omega^8 & B\omega^9 & G\omega^{10} & -B\omega^{11} & -G\omega^{12} & \cdots \\ \vdots & \vdots & \vdots & \vdots & \vdots & \vdots & \vdots & \ddots \end{bmatrix}$$

$$
BA = (AB)^T =
\begin{bmatrix}
-G & -B\omega & G\omega^2 & B\omega^3 & -G\omega^4 & -B\omega^5 & G\omega^6 & \cdots \\
B\omega & -G\omega^2 & -B\omega^3 & G\omega^4 & B\omega^5 & -G\omega^6 & -B\omega^7 & \cdots \\
G\omega^2 & B\omega^3 & -G\omega^4 & -B\omega^5 & G\omega^6 & B\omega^7 & -G\omega^8 & \cdots \\
-B\omega^3 & G\omega^4 & B\omega^5 & -G\omega^6 & -B\omega^7 & G\omega^8 & B\omega^9 & \cdots \\
-G\omega^4 & -B\omega^5 & G\omega^6 & B\omega^7 & -G\omega^8 & -B\omega^9 & G\omega^{10} & \cdots \\
B\omega^5 & -G\omega^6 & -B\omega^7 & G\omega^8 & B\omega^9 & -G\omega^{10} & -B\omega^{11} & \cdots \\
G\omega^6 & B\omega^7 & -G\omega^8 & -B\omega^9 & G\omega^{10} & B\omega^{11} & -G\omega^{12} & \cdots \\
\vdots & \vdots & \vdots & \vdots & \vdots & \vdots & \vdots & \ddots
\end{bmatrix}
$$

$$
BB =
\begin{bmatrix}
1 & & -\omega^2 & & \omega^4 & & -\omega^6 & \\
& \omega^2 & & -\omega^4 & & \omega^6 & & \cdots \\
-\omega^2 & & \omega^4 & & -\omega^6 & & \omega^8 & \\
& -\omega^4 & & \omega^6 & & -\omega^8 & & \cdots \\
\omega^4 & & -\omega^6 & & \omega^8 & & -\omega^{10} & \\
& \omega^6 & & -\omega^8 & & \omega^{10} & & \cdots \\
-\omega^6 & & \omega^8 & & -\omega^{10} & & \omega^{12} & \\
\vdots & & \vdots & & \vdots & & & \ddots
\end{bmatrix}
$$

For simplicity, $G(\omega)$ and $B(\omega)$ are indicated as G and B, respectively. Also, the Σ symbol is omitted. Only the matrix elements enclosed by the dotted line remain and agree with (3.41) by setting $a_0 = 1$, $a_3 \sim a_6 = 0$, $b_0 = 0$ and $b_2 \sim b_6 = 0$ in (3.43). Also, since the matrix elements have regularity, expansion of further orders will be easy.

Figure 3.33 shows the results of modeling by changing the order of the real rational function to the same measurement result as Figure 3.31. If the order is too low, the modeled result does not trace the measurement result. Also, if the order is too high, the modeled result picks up measurement noise. From this, it is necessary for device modeling to set an appropriate order considering the physical structure.

3.3.2 *MOSFET parasitic resistances at millimeter wave [1]*

Predictive circuit simulation is possible only if device models that accurately reproduce measurement results are available. To build a reliable MOSFET model, parasitic resistances in device layout, shown in Figure 3.34, have to be estimated in terahertz region. The parasitic resistances at terahertz frequencies are generally different from dc values due to non-quasi static effects. To extract the parasitic resistances, "cold-bias de-embedding" or the "cold-FET method" [33] is applied.

Theoretically, since the channel resistance is proportional to the reciprocal of the overdrive voltage (gate voltage minus threshold voltage), parasitic resistances can be extracted by mathematically zeroing the channel resistance by extrapolating the overdrive voltage to infinity. This is the essence of cold-bias de-embedding. In RF cold-bias de-embedding, real parts of the impedance matrix of a MOSFET at

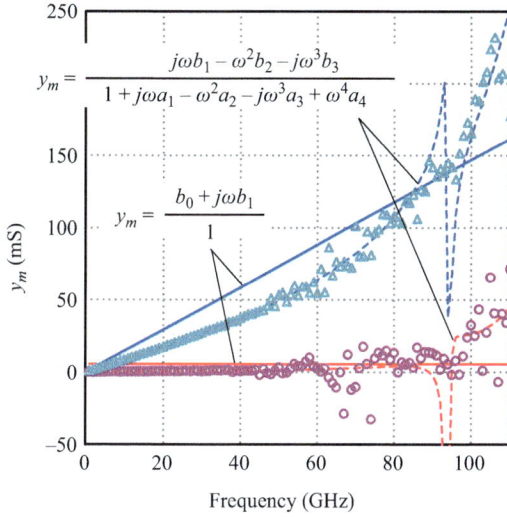

$$y_m = \frac{j\omega b_1 - \omega^2 b_2 - j\omega^3 b_3}{1 + j\omega a_1 - \omega^2 a_2 - j\omega^3 a_3 + \omega^4 a_4}$$

$$y_m = \frac{b_0 + j\omega b_1}{1}$$

Figure 3.33 *Comparison between results modeled using real rational function and measurement results (y_{in} and y_{out}). Results are shown by S parameters*

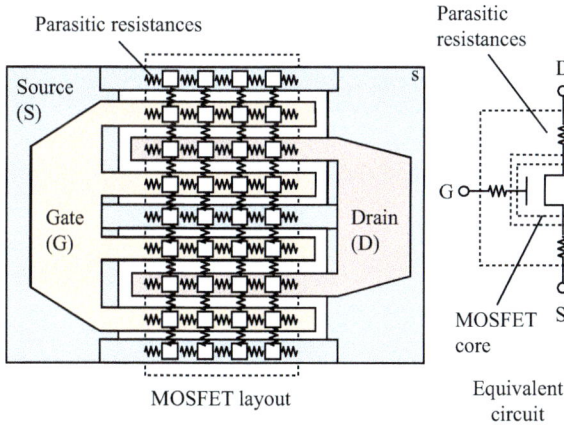

Figure 3.34 *Schematic layout of a MOSFET and its equivalent circuit highlighting parasitic resistances. Copyright 2015 IEICE. Reproduced, with permission, from [1]*

a low frequency is used. However, since real measurement data show dispersion, we proposed a new extraction method [34]. In this method, measured data up to 110 GHz were first smoothed out as shown in Figure 3.35 before further processing. Figure 3.36 shows comparison of extracted resistances with and without smoothing as a function of the gate voltage. Clearly, smoothing in the previous step has given more well-behaved results. Now the extracted resistance ($R_D + R_{ch} + R_S$) is expected to be a linear function of the reciprocal overdrive voltage $(V_{GS} - V_{th})^{-1}$. But in practice, that turned out not to be the case, if the value of V_{th} was equal to the dc value of 0.4 V, as shown in Figure 3.37. We, therefore, introduced the mmw threshold voltage (mmw V_{th}), equal to 0.5 V, which makes $R_D + R_{ch} + R_S$ a linear function of $(V_{GS} - V_{th})^{-1}$. The parasitic resistance $R_D + R_S$ can then be extracted reliably by linear extrapolation $(V_{GS} - V_{th})^{-1} \to 0$, as shown in Figure 3.37, which amounts to bringing R_{ch} to zero. After reliably extracting the parasitic resistances, we successfully developed

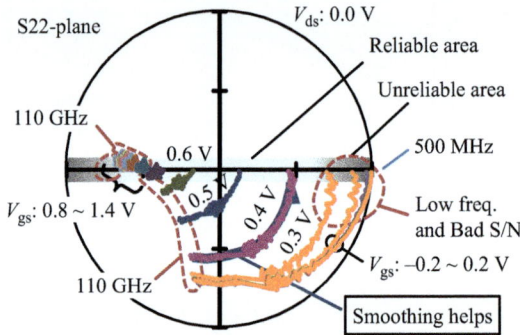

Figure 3.35   Measured S_{22} at 0.5–110 GHz under different bias conditions. Immittances that are located near the center of the reflection-coefficient plane (Smith chart) are more reliable and less susceptible to measurement uncertainties than those located far from the center because of the mathematical form of the linear fractional transformation that relates reflection coefficients to immittances. S_{22} at 500 MHz are in the unreliable region. Copyright 2015 IEICE. Reproduced, with permission, from [1]

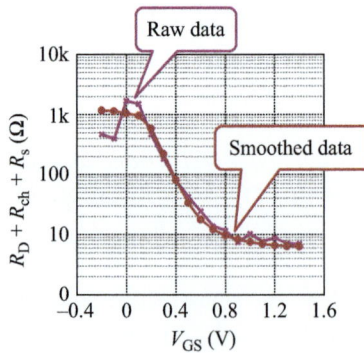

Figure 3.36   Channel resistance (R_{ch}) plus parasitic resistances ($R_D + R_S$) extracted from raw and smoothed data at 500 MHz versus the gate voltage V_{GS}. Copyright 2015 IEICE. Reproduced, with permission, from [1]

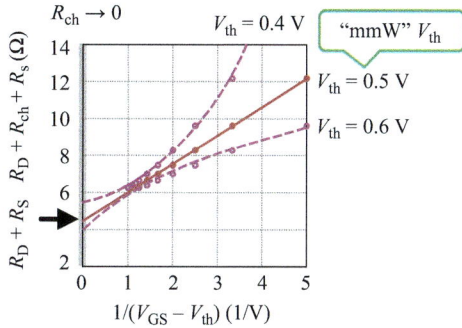

Figure 3.37 *Extraction of parasitic resistance $R_D + R_S$ by linear extrapolation ($V_{GS} - V_{th})^{-1} \to 0$ works only if the value of V_{th} is the millimeter-wave value of 0.5 V. The dc threshold voltage is 0.4 V. Copyright 2015 IEICE. Reproduced, with permission, from [1]*

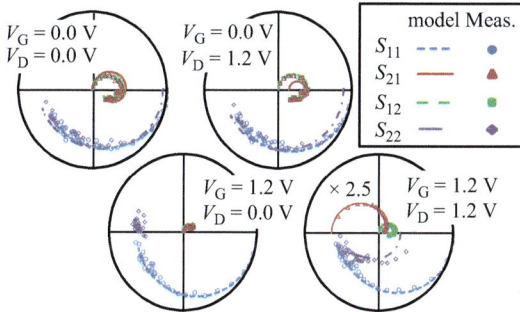

Figure 3.38 *S parameters of a 65-nm NMOSFET at 0.5–330 GHz (not including 170–220 GHz). Copyright 2015 IEICE. Reproduced, with permission, from [1]*

a MOSFET model applicable to 300-GHz nonlinear simulation. Comparison of *S* parameters obtained from simulation with the new model and measurement is shown in Figure 3.38. The agreement is reasonable throughout the drain and gate voltage ranges of 0–1.2 V.

References

[1] M. Fujishima, S. Amakawa, K. Takano, K. Katayama, and T. Yoshida, "Terahertz CMOS design for low-power and high-speed wireless communication," *IEICE Transactions Electronics*, vol. E98-C, no. 12, pp. 1091–1104, 2015.

[2] Y. Manzawa, Y. Goto, M. Fujishima, "Bond-based design for MMW CMOS circuit optimization," *Asia-Pacific Microwave Conference*, pp. 1–4, Dec. 2008.

[3]   R. Fujimoto, M. Motoyoshi, K. Takano, M. Fujishima, "A 120 GHz/140 GHz dual-channel OOK receiver using 65 nm CMOS technology," *IEICE Transactions on Fundamentals*, vol. E96-A, no. 2, pp. 486–493, 2013.

[4]   H. W. Ott, *Electromagnetic Compatibility Engineering*, Wiley, New York, 2009.

[5]   Y. Manzawa, M. Sasaki, M. Fujishima, "High-attenuation power line for wideband decoupling," *IEICE Transactions on Electronics*, vol. E92-C, no. 6, pp. 792–797, 2009.

[6]   R. B. Adler, L. J. Chu, R. M. Fano, *Electromagnetic Energy Transmission and Radiation*, Wiley, New York, 1960.

[7]   Y. Natsukari, M. Fujishima, "36mW 63GHz CMOS differential low-noise amplifier with 14GHz bandwidth," *Symposium on VLSI Circuits*, pp. 252–253, Jun. 2009.

[8]   R. Goda, S. Amakawa, K. Katayama, K. Takano, T. Yoshida, M. Fujishima, "Characterization of wideband decoupling power line with extremely low characteristic impedance for millimeter-wave CMOS circuits," *International Conference on Microelectronic Test Structures*, pp. 220–223, Mar. 2015.

[9]   S. Amakawa, R. Goda, K. Katayama, K. Takano, T. Yoshida, M. Fujishima, "Wideband CMOS decoupling power line for millimeter-wave applications," *IEEE MTT-S International Microwave Symposium*, pp. 1–4, May 2015.

[10]   N. Fajri, R. Minami, K. Okada, A. Matsuzawa, "A new structure of low-loss MIM transmission line for 60 GHz," *IEICE Society Conference*, C-12-12, Sep. 2012 (in Japanese).

[11]   G. Tretter, D. Fritsche, J. D. Leufker, C. Carta, F. Ellinger, "Zero-Ohm transmission lines for millimetre-wave circuits in 28 nm digital CMOS," *Electronics Letters*, vol. 51, no. 11, pp. 845–847, 2015.

[12]   M. Fujishima, M. Motoyoshi, K. Katayama, K. Takano, N. Ono, and R. Fujimoto, "98 mW 10Gbps wireless transceiver chipset with D-band CMOS circuits," *IEEE Journal of Solid-State Circuits*, vol. 48, no. 10, pp. 2273–2284, 2013.

[13]   K. H. K. Yau, E. Dacquay, I. Sarkas, S. P. Voinigescu, "Device and IC characterization above 100 GHz," *IEEE Microwave Magazine*, vol. 13, no. 1, pp. 30–54, 2012.

[14]   R. B. Marks, "A multiline method of network analyzer calibration," *IEEE Transactions on Microwave Theory and Techniques*, vol. 39, no. 7, pp. 1205–1215, 1991.

[15]   A. Orii, M. Suizu, S. Amakawa, *et al.,* "On the length of thru standard for TRL de-embedding on Si substrate above 110 GHz," *International Conference on Microelectronic Test Structures*, pp. 81–86, Mar. 2013.

[16]   K. H. K. Yau, I. Sarkas, A. Tomkins, P. Chevalier, S. P. Voinigescu, "On-wafer S-parameter de-embedding of silicon active and passive devices up to 170 GHz," *IEEE MTT-S International Microwave Symposium*, pp. 600–603, May 2010.

[17]   D. F. Williams, P. Corson, J. Sharma, *et al.*, "Calibration-kit design for millimeter-wave silicon integrated circuits," *IEEE Transactions on Microwave Theory and Techniques*, vol. 61, no. 7, pp. 2685–2694, 2013.

[18] F. J. Schmückle, R. Doerner, G. N. Phung, W. Heinrich, D. Williams, U. Arz, "Radiation, multimode propagation, and substrate modes in W-band CPW calibrations," *European Microwave Conference*, pp. 297–300, Oct. 2011.

[19] S. Amakawa, K. Katayama, K. Takano, T. Yoshida, and M. Fujishima, 'Comparative Analysis of On-Chip Transmission Line De-Embedding Techniques," *2015 IEEE International Symposium on Radio-Frequency Integration Technology*, pp. 91–93, 2015.

[20] D. F. Williams, F.-J. Schmückle, R. Doerner, G. N. Phung, U. Arz, W. Heinrich, "Crosstalk corrections for coplanar-waveguide scattering-parameter calibrations," *IEEE Transactions on Microwave Theory and Techniques*, vol. 62, no. 8, pp. 1748–1761, 2014.

[21] S. Amakawa, A. Orii, K. Katayama, *et al.*, "Design of well-behaved low-loss millimetre-wave CMOS transmission lines," *IEEE Workshop on Signal and Power Integrity*, pp. 1–4, May 2014.

[22] D. F. Williams, R. B. Marks, "Transmission line capacitance measurement," *IEEE Microwave and Guided Wave Letters*, vol. 1, no. 9, pp. 243–245, 1991.

[23] R. B. Marks, D. F. Williams, "Characteristic impedance determination using propagation constant measurement," *IEEE Microwave and Guided Wave Letters*, vol. 1, no. 6, pp. 141–143, 1991.

[24] S. Amakawa, K. Takano, K. Katayama, M. Motoyoshi, T. Yoshida, M. Fujishima, "On the choice of cascade de-embedding methods for on-wafer S-parameter measurement," *IEEE International Symposium on Radio-Frequency Integration Technology*, pp. 137–139, Nov. 2012.

[25] A. M. Mangan, S. P. Voinigescu, M.-T. Yang, M. Tazlauanu, "De-embedding transmission line measurements for accurate modeling of IC designs," *IEEE Transactions on Electron Devices*, vol. 53, no. 2, pp. 235–241, 2006.

[26] T. Sekiguchi, S. Amakawa, N. Ishihara, K. Masu, "On the validity of bisection-based thru-only de-embedding," *International Conference on Microelectronic Test Structures*, pp. 66–71, Mar. 2010.

[27] D. F. Williams, U. Arz, H. Grabinski, "Characteristic-impedance measurement error on lossy substrates," *IEEE Microwave and Wireless Components Letters*, vol. 11, no. 7, pp. 299–301, 2001.

[28] K. Katayama, M. Motoyoshi, K. Takano, C. Y. Li, S. Amakawa, M. Fujishima, "E-band 65nm CMOS low-noise amplifier design using gain-boost technique," *IEICE Transactions on Electronics*, vol. E97-C, no. 6, pp. 476–485, 2014.

[29] K. Takano, S. Amakawa, K. Katayama, M. Motoyoshi, M. Fujishima, "Characteristic impedance determination technique for CMOS on-wafer transmission line with large substrate loss," *79th ARFTG Conference*, pp. 1–4, Jun. 2012.

[30] K. Takano, K. Katayama, S. Mizukusa, S. Amakawa, T. Yoshida, M. Fujishima, "Systematic calibration procedure of process parameters for electromagnetic field analysis of millimeter-wave CMOS devices," *International Conference on Microelectronic Test Structures* , pp. 1–5, Mar. 2015.

[31]   K. Takano, K. Katayama, T. Yoshida, *et al.*, "Calibration of process parameters for electromagnetic field analysis of CMOS devices up to 330 GHz," *International Symposium on Radio-Frequency Integration Technology*, pp. 94–96, 2015.

[32]   K. Takano, S. Amakawa, K. Katayama, M. Motoyoshi, M. Fujishima, "Modeling of short-millimeter-wave CMOS transmission line with lossy dielectrics with specific absorption spectrum," *IEICE Transactions on Electronics*, vol. E96-C, no. 10, pp. 1311–1318, 2013.

[33]   P. H. Aaen, J. A. Plá, J. Wood, *Modeling and Characterization of RF and Microwave Power FETs*, Cambridge University Press, Cambridge, 2007.

[34]   K. Katayama, S. Amakawa, K. Takano, M. Fujishima, "300-GHz MOSFET model extracted by an accurate cold-bias de-embedding technique," *IEEE MTT-S International Microwave Symposium*, pp. 1–4, May 2015.

Chapter 4
Transceiver design

4.1 Transmitter

In this section, we discuss a 300-GHz complementary metal-oxide-semiconductor (CMOS) transmitter (TX) reported previously in [1]. It covers the currently unallocated frequency range from 275 to 305 GHz (Figure 4.1) with six 5-GHz-wide channels, as shown in Figure 4.2. The overall bandwidth is more than three times the 60-GHz unlicensed bandwidth of 9 GHz. According to the process design kit provided by the foundry, NMOS f_{max} for the 40-nm CMOS process that we used is about 280 GHz. The measured f_{max} values were somewhat lower. Our TX, therefore, is an above-f_{max} TX. While quadrature amplitude modulation (QAM) signal transmission at 300 GHz or above has already been reported [2,3], achieving this using a CMOS technology has been a tremendous challenge, as will be discussed in the next section. Our TX is capable of 32QAM 17.5-Gbit/s/ch signal transmission.

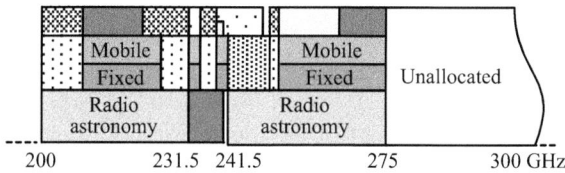

Figure 4.1 Frequency allocations above 200 GHz in the United States. Copyright 2016 IEEE. Reproduced, with permission, from [1]

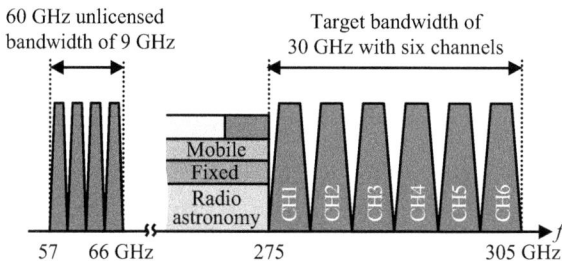

Figure 4.2 Target band plan and channel allocation for 300-GHz ultrahigh-speed wireless communication. Copyright 2016 IEEE. Reproduced, with permission, from [1]

In Section 4.1.1, we discuss the choice of the architecture in view of the transistor f_{max} and clarify the technical challenges faced while building an above-f_{max} TX. Section 4.1.2 explains the designs of the constituent blocks and the overall TX. Section 4.1.3 analyzes the characteristics of the key enabling component, a highly linear subharmonic mixer called a *cubic mixer*. Section 4.1.4 presents the measured performance of the TX. Finally, Section 4.1.5 discusses the doubler-based architecture to generate high-output power.

4.1.1 Architectural consideration [1]

Figure 4.3 shows possible architectures of terahertz TXs based on state-of-the-art above-200 GHz TXs. The choice of the architecture strongly depends on the transistor f_{max}. Above f_{max}, the transistor gain falls below unity and the transistor becomes passive. If f_{max} is sufficiently higher than the transmitted signal frequency, radio frequency (RF), as is typical when using InP technology (Figure 4.3, first row), the ordinary power amplifier (PA)-last architecture [4–7] is the most suitable. In this architecture, the intermediate-frequency (IF) signal is upconverted to RF by a mixer and the signal is amplified at RF. Complex modulation formats such as QAM can be used, provided the signal path has sufficient linearity. Since f_{max} should be very high when this architecture is adopted, the conversion gain of the mixer should also be reasonably high. The presence of a PA is not a requirement [8,9].

If f_{max} is comparable to or lower than RF, as is typical when using CMOS technology, a PA-less architecture must be adopted. State-of-the-art CMOS TXs adopt a tripler-last [10] (Figure 4.3, second row) or quadrupler-last architecture [11]. A drawback of these architectures is their low spectral efficiency due to signal bandwidth spreading caused by frequency tripling or quadrupling. Another drawback is that the use of multibit digital modulation is very difficult, if not impossible. An exception to this is quaternary phase-shift keying (QPSK) combined with frequency tripling.

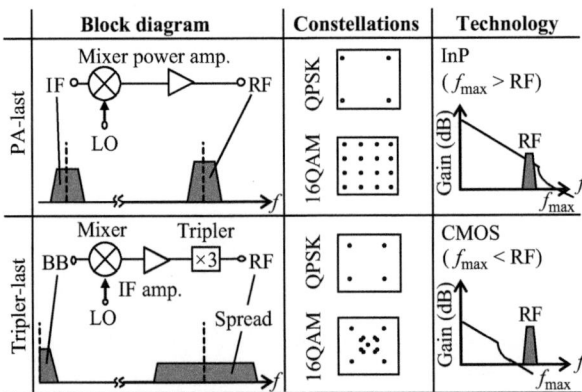

Figure 4.3   *Possible architectures of terahertz transmitters and corresponding output power spectra, output signal constellations, and frequency dependence of transistor gain*

When a QPSK-modulated IF signal undergoes frequency tripling, the resulting signal constellation remains that of QPSK with some symbol permutation. Such a tripler-last QPSK TX has been reported [12]. However, a 16QAM constellation, for example, would suffer severe distortion upon frequency tripling (Figure 4.3, second row). If the 300-GHz band is to be seriously considered as a platform for ultrahigh-speed wireless communication, QAM capability will be a requisite, because a wide bandwidth alone is not sufficient for achieving very high data rates.

Figure 4.4 shows a possible PA-less, mixer-last architecture that has the PA removed from the PA-last architecture (Figure 4.3, first row). In this configuration, signal bandwidth spreading should not occur and QAM should work, except that the amplitude will be much smaller than that in the PA-last case. However, no PA-less, mixer-last CMOS TXs operating above f_{max} appear to have been reported, presumably because the layout would become too complicated, as shown in the schematic diagram of Figure 4.4. Our goal here is to develop a 300-GHz CMOS TX operating above f_{max}. The output power from a single mixer is inevitably very low. That means that massively parallel power combining is necessary at RF. However, if we attempted to achieve this using ordinary three-port mixers, as in Figure 4.4, the layout would become extremely complicated with a large number of crossovers. The mixers would typically be double-balanced mixers, thus doubling the number of wires. It would be extremely difficult to achieve the correct phase relationship between different branches. In contrast, the tripler-last architecture (Figure 4.3, second row) allows massive power combining with a much simpler layout because it does not require local-oscillator (LO) signal paths.

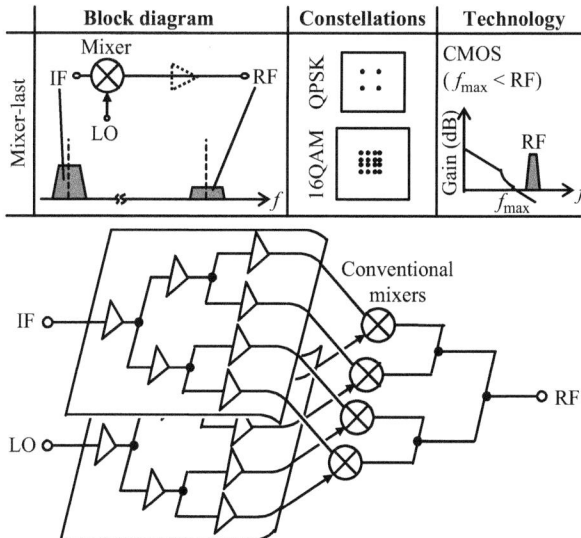

Figure 4.4   Easily conceivable mixer-last architecture. It is difficult to realize in practice because the layout would become extremely complicated with a large number of crossovers. Copyright 2016 IEEE. Reproduced, with permission, from [1]

To summarize, the main technical challenges faced when building an above-f_{max} PA-less TX in CMOS are as follows: (i) giving the QAM capability required to achieve high data rates. That means that frequency multipliers should not be used because they badly distort QAM signals. (ii) Performing massive power combining to obtain sufficient output power without undue layout complication and/or die area explosion.

We reconcile these conflicting requirements by introducing a mixer-last architecture that employs a tripler-based, highly linear subharmonic mixer, which we call the *cubic mixer* (Figure 4.5). Since it is a mixer, not a tripler, the TX is QAM-capable and there is no signal bandwidth spreading. The layout of the cubic mixer is as simple as that of a tripler, and this simplicity is vital for realizing massive power combining without undue layout complication.

4.1.2   Circuit design [1]

As shown in Figure 4.5, the 300-GHz CMOS TX consists of a double-balanced mixer, a power splitter, cubic mixers, and a power combiner. Since the cubic mixer is the key component of this TX, we start this section with a description of it. We then introduce the power splitter and power combiner. Finally, we explain the overall design of the TX.

4.1.2.1   Cubic mixer

The cubic mixer is used to upconvert a digitally modulated IF signal at IF_2 to RF with high linearity (Figure 4.5). As the name suggests, it uses the cubic nonlinearity of the MOSFET and is essentially a tripler. It receives a two-tone-like signal composed of modulated IF_2 and pure LO. When the two-tone-like input is cubed by a tripler, first-to third-order signals are generated in accordance with

$$(LO + IF_2)^3 = LO^3 + 3LO^2 \cdot IF_2 + 3LO \cdot IF_2^2 + IF_2^3 \tag{4.1}$$

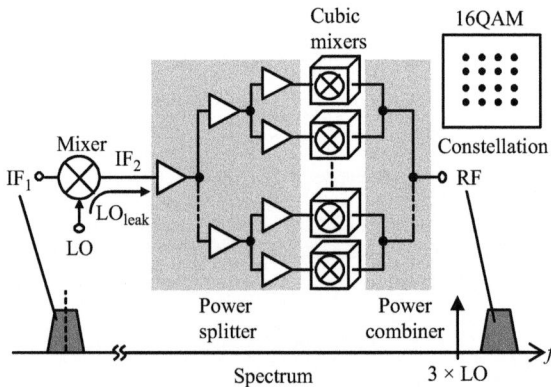

Figure 4.5   *QAM-capable cubic-mixer-last architecture. Its layout is as simple as that of the tripler-last architecture. There is no bandwidth spreading at RF. Copyright 2016 IEEE. Reproduced, with permission, from [1]*

as illustrated in Figure 4.6. To be exact,

$$(\cos \omega_{LO} t + \cos \omega_{IF2} t)^3$$

$$= \frac{3}{4} [3 \cos \omega_{LO} t + \cos (2\omega_{LO} - \omega_{IF2})t + \cos (\omega_{LO} - 2\omega_{IF2})t + 3 \cos \omega_{IF2} t]$$

$$+ \frac{1}{4} [\cos 3\omega_{LO} t + 3 \cos (2\omega_{LO} + \omega_{IF2})t + 3 \cos (\omega_{LO} + 2\omega_{IF2})t + \cos 3\omega_{IF2} t].$$

$$(4.2)$$

The last four terms of (4.2) are the high-frequency (near $3\omega_{LO}$) components and are shown in Figure 4.6. By appropriately tuning the power levels of LO and IF_2, the second subharmonic mixing term, $3LO^2 \cdot IF_2$ in (4.1), becomes the dominant output. We use this term for linear signal upconversion. The optimal LO-power-to-IF_2-power ratio is $P_{LO}/P_{IF2} = 2$ [13]. Figure 4.7 shows a schematic of the cubic mixer, which

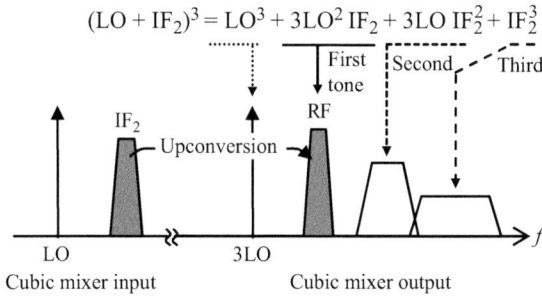

Figure 4.6   *Operation principle of the cubic mixer. The second subharmonic mixing term, $3 LO^2 \cdot IF_2$, is used for linear upconversion. $LO < IF_2$ in this example. Copyright 2016 IEEE. Reproduced, with permission, from [1]*

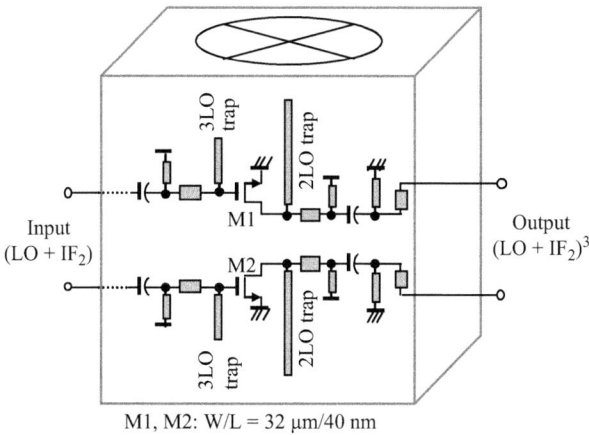

M1, M2: W/L = 32 μm/40 nm

Figure 4.7   *Schematic diagram of the cubic mixer. Copyright 2016 IEEE. Reproduced, with permission, from [1]*

LO = 97 GHz, 5 dB m, IF$_2$ = 103 GHz

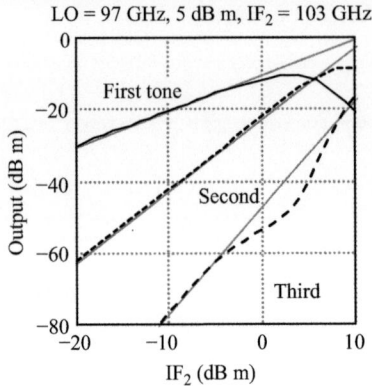

Figure 4.8   Simulated output power of the cubic mixer as a function of input IF$_2$ power. Copyright 2016 IEEE. Reproduced, with permission, from [1]

LO = 97 GHz, 5 dB m, IF$_2$ = −5 dB m

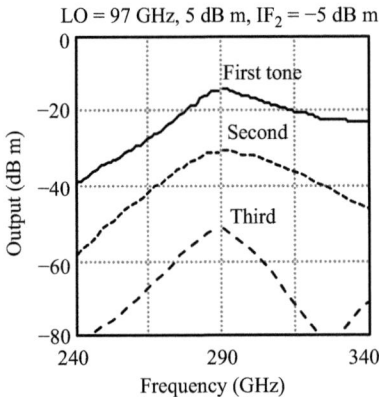

Figure 4.9   Simulated output power of the cubic mixer as a function of frequency. Copyright 2016 IEEE. Reproduced, with permission, from [1]

is a tripler with a differential input and output. As shown in the circuit simulation result in Figure 4.8, upconversion is accomplished with good linearity. The desired first tone is the dominant output at frequencies of interest, as shown in Figure 4.9.

A more detailed analysis of the frequency-dependent characteristics of the cubic mixer is given in Section 4.1.3.

4.1.2.2   Power splitter

Since the NMOS f_{max} of the 40-nm CMOS process that we used is at most 280 GHz, the cubic mixer acts as a passive mixer with a rather low conversion gain. Massively parallel amplification at IF$_2$ and power combining at RF are required, and we perform 32-way power combining.

An orthodox approach to power splitting and amplification without introducing layout crossovers is to use passive baluns and differential amplifiers [5]. However,

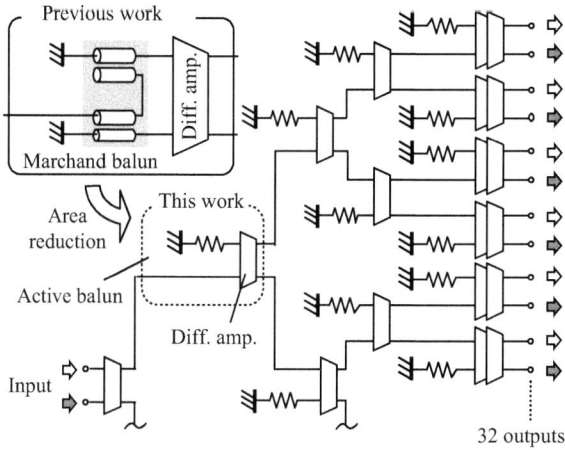

Figure 4.10   32-way power-splitting network composed of differential-amplifier-based active baluns. The area is significantly smaller than a passive-balun-based network [5]. Copyright 2016 IEEE. Reproduced, with permission, from [1]

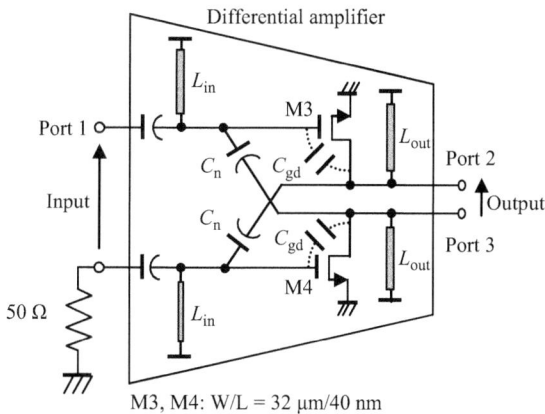

M3, M4: W/L = 32 μm/40 nm

Figure 4.11   Schematic diagram of the active balun. Copyright 2016 IEEE. Reproduced, with permission, from [1]

passive baluns operating at 100 GHz occupy a large area, and the area required for 32-way power splitting would become excessive if we adopted a passive-balun-based network. As shown in Figure 4.10, we perform crossover-free 32-way power splitting using differential-amplifier-based active baluns with single-ended feeding.

Figure 4.11 shows a schematic of the active balun. It is a capacitively neutralized differential amplifier with a single-ended input and a differential output. This type of balun is known to have some limitations compared with passive baluns, including the inherent imbalance between the two outputs [14]. However, it can be designed to perform reasonably well over a not-so-wide bandwidth.

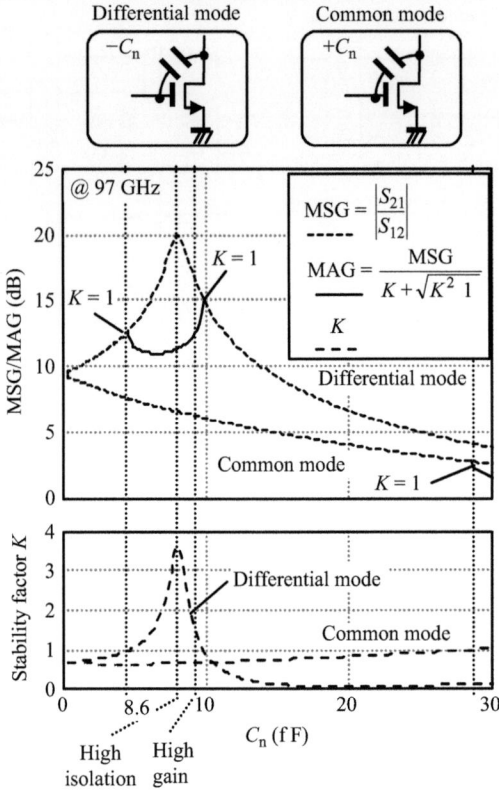

Figure 4.12 *Differential- and common-mode half-circuit characteristics of the active balun (Figure 4.11) at 97 GHz as a function of feedback (neutralization) capacitance C_n. The upper graph shows the maximum stable gain (MSG) and MAG. The lower graph shows the stability factor K. Copyright 2016 IEEE. Reproduced, with permission, from [1]*

Figure 4.12 shows the differential- and common-mode half circuits of the core amplifier. The capacitor C_n functions as a neutralizing capacitor only in the differential mode, and it lowers the gain in the common mode. Figure 4.12 also shows the gain and the stability factor K [15] of the half circuits as a function of the capacitance C_n. If the design goal is high gain, a value of C_n could be chosen from an *over-neutralized region* [5], in which the maximum available gain (MAG) [15] has a local maximum. The interrelationship between the MAG and K can be visualized, and the design space be further explored using a graphical chart called the *MAG-K chart*, discussed in Section 2.3 [16,17]. The graphs in Figure 4.12 can be understood as cross-sectional gain and K profiles along the Y (shunt–shunt)-feedback circle on a MAG-K chart (Section 2.3) [20]. Here we choose a value of C_n such that the highest reverse isolation is achieved (within the limited design space spanned only by C_n), because we must be able to tune L_{in} and L_{out} in Figure 4.11 independently to optimize

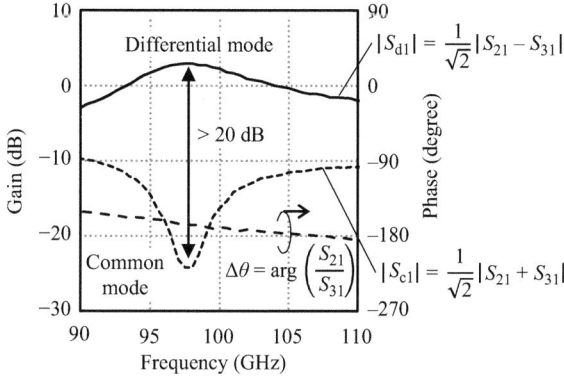

Figure 4.13   Simulated differential-mode gain $|S_{d1}|$, common-mode gain $|S_{c1}|$, and phase balance $\Delta\theta$ for the active balun (Figure 4.11). Copyright 2016 IEEE. Reproduced, with permission, from [1]

the balun characteristics. Figure 4.13 shows the frequency-dependent characteristics of the active balun. Thanks to the high isolation between the input and output ports, we can select the lengths of the shunt stubs implementing L_{in} and L_{out} independently to obtain good input and output matching. We tuned the stub lengths such that the phase difference between the output ports is $\Delta\theta \simeq \arg(S_{21}/S_{31}) = 180°$ at around 100 GHz. The common-mode rejection ratio is maximized near this frequency as shown in Figure 4.13.

4.1.2.3   Power combiner

We performed 32-way power splitting at IF_2 and fed 16 cubic mixers differentially. We must therefore perform 32-way power combining at RF. We introduce a passive eight-way combiner called the *quad-rat-race*, shown in Figure 4.14, and use four quad-rat-races to perform 32-way power combining. The quad-rat-race is an extension of the ordinary rat-race balun [18], shown in Figure 4.15(a). The circumference of the rat race is 1.5λ. According to circuit theory, if the access transmission lines have a characteristic resistance [19–21] of R_0, the line constituting the ring should have a characteristic resistance of $\sqrt{2}R_0$ [18]. Then, all the diagonal elements of the S-matrix of the rat-race balun become zero ($S_{ii} = 0$), as shown in Figure 4.15(a). This can be understood by recognizing that port 3 (port 2) becomes a virtual ground node when port 2 (port 3) is the driving port and that each of the λ/4-sections of the ring serves as a λ/4-transformer.

The quad-rat-race has four differential input ports or eight single-ended input ports, as shown in Figure 4.16(a). It performs differential-to-single-ended conversion and eight-way power combining. A differential input port consists of a pair of single-ended ports. These input ports are placed λ/2 apart from each other. The circumference of the ring is 4.5λ. A signal injected into a single-ended port splits into clockwise and counterclockwise traveling waves, each having half the input power. The first (second) graph in Figure 4.16(b) shows the traveling-wave voltages along the ring

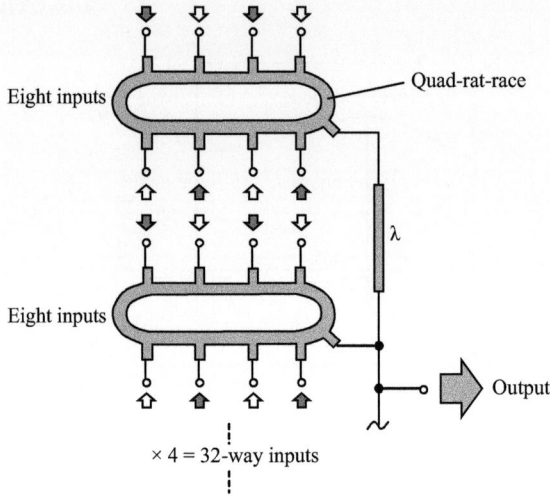

Figure 4.14   *32-way power combining network consisting of four quad-rat-races and two sections of transmission line of length λ. Copyright 2016 IEEE. Reproduced, with permission, from [1]*

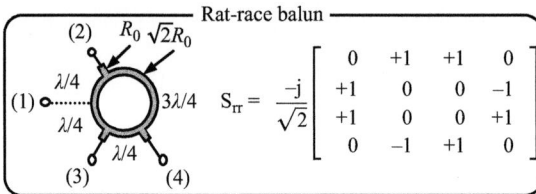

(a)

$$
\begin{bmatrix} b_1 \\ b_2 \\ b_3 \\ b_4 \\ b_5 \\ b_6 \\ b_7 \\ b_8 \\ b_9 \\ b_{10} \end{bmatrix} = \frac{-j}{\sqrt{8}}
\begin{bmatrix}
0 & +1 & -1 & +1 & -1 & +1 & -1 & +1 & -1 & 0 \\
+1 & -\frac{3j}{\sqrt2} & -\frac{j}{\sqrt2} & +\frac{j}{\sqrt2} & -\frac{j}{\sqrt2} & 0 & 0 & 0 & 0 & -1 \\
-1 & -\frac{j}{\sqrt2} & -\frac{3j}{\sqrt2} & -\frac{j}{\sqrt2} & +\frac{j}{\sqrt2} & 0 & 0 & 0 & 0 & +1 \\
+1 & +\frac{j}{\sqrt2} & -\frac{j}{\sqrt2} & -\frac{3j}{\sqrt2} & -\frac{j}{\sqrt2} & 0 & 0 & 0 & 0 & -1 \\
-1 & -\frac{j}{\sqrt2} & +\frac{j}{\sqrt2} & -\frac{j}{\sqrt2} & -\frac{3j}{\sqrt2} & 0 & 0 & 0 & 0 & +1 \\
+1 & 0 & 0 & 0 & 0 & -\frac{3j}{\sqrt2} & -\frac{j}{\sqrt2} & +\frac{j}{\sqrt2} & -\frac{j}{\sqrt2} & +1 \\
-1 & 0 & 0 & 0 & 0 & -\frac{j}{\sqrt2} & -\frac{3j}{\sqrt2} & -\frac{j}{\sqrt2} & +\frac{j}{\sqrt2} & -1 \\
+1 & 0 & 0 & 0 & 0 & +\frac{j}{\sqrt2} & -\frac{j}{\sqrt2} & -\frac{3j}{\sqrt2} & -\frac{j}{\sqrt2} & +1 \\
-1 & 0 & 0 & 0 & 0 & -\frac{j}{\sqrt2} & +\frac{j}{\sqrt2} & -\frac{j}{\sqrt2} & -\frac{3j}{\sqrt2} & -1 \\
0 & -1 & +1 & -1 & +1 & +1 & -1 & +1 & -1 & 0
\end{bmatrix}
\begin{bmatrix} a_1 \\ a_2 \\ a_3 \\ a_4 \\ a_5 \\ a_6 \\ a_7 \\ a_8 \\ a_9 \\ a_{10} \end{bmatrix}
$$

(b)

Figure 4.15   *(a) Rat-race balun and its S-matrix in the ideal lossless case. R_0 is the characteristic resistance of the access line. Port numbers are indicated in parentheses. (b) S-matrix of the quad-rat-race in an ideal case, where the ring has a characteristic resistance of $R_0/\sqrt{2}$. Port numbers are shown in Figure 4.16(a). Copyright 2016 IEEE. Reproduced, with permission, from [1]*

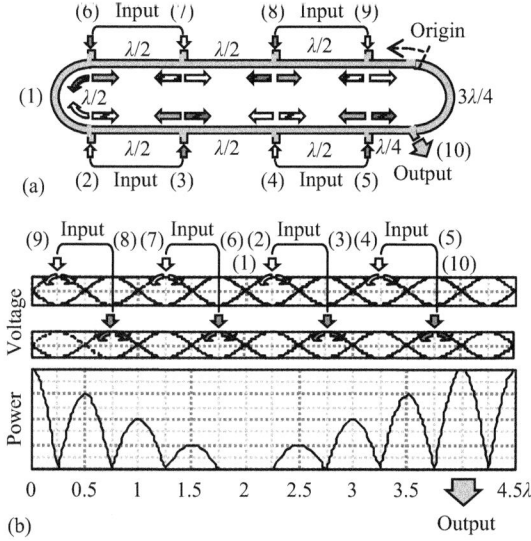

Figure 4.16 (a) *Quad-rat-race. Port numbers are indicated in parentheses.*
(b) *Traveling-wave voltages and the power available along the ring.*
Copyright 2016 IEEE. Reproduced, with permission, from [1]

originating from the positive (negative) ports. A pair of counter-rotating traveling waves originating from an input port cancel each other when they reach other input ports, that is, the input ports become virtual ground nodes. On the other hand, a pair of counter-rotating traveling waves are superposed constructively at the output port. The power available along the ring is plotted in the bottom graph of Figure 4.16(b). Altogether, 16 traveling waves are summed at the output port and eight-way power combining is achieved.

Figure 4.15(b) shows the S-matrix of a quad-rat-race when the characteristic resistance of the ring is $R_0/\sqrt{2}$. If the inputs to the quad-rat-race are four ideally differential signals, the S-matrix can be simplified by the following substitution:

$$\begin{cases} a_2' = (a_2 - a_3 + a_4 - a_5)/2, \\ a_6' = (a_6 - a_7 + a_8 - a_9)/2, \\ b_2' = (b_2 - b_3 + b_4 - b_5)/2, \\ b_6' = (b_6 - b_7 + b_8 - b_9)/2. \end{cases} \tag{4.3}$$

The result is

$$\begin{bmatrix} b_1 \\ b_2' \\ b_6' \\ b_{10} \end{bmatrix} = \mathbf{S}_{rr} \begin{bmatrix} a_1 \\ a_2' \\ a_6' \\ a_{10} \end{bmatrix}, \tag{4.4}$$

where $\mathbf{S}_{rr}$ is the S-matrix of the rat race given in Figure 4.15(a). Equation (4.4) clearly shows that the quad-rat-race is an extension of the rat race.

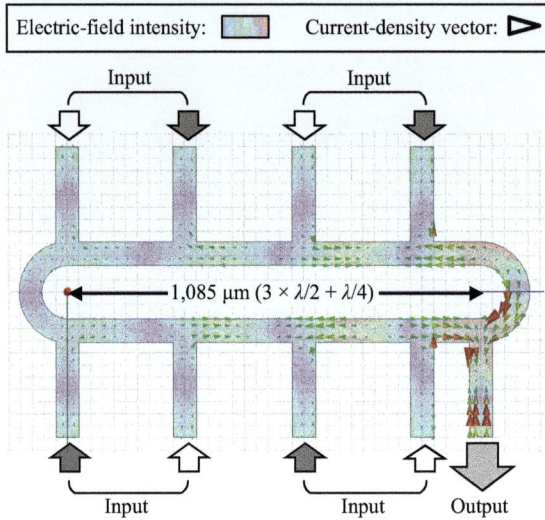

Figure 4.17   Simulated electric-field intensity and current-density vectors in a lossless quad-rat-race having a simpler layout than the actual design (in Figure 4.20), provided as an illustration of the basic operation. Copyright 2016 IEEE. Reproduced, with permission, from [1]

Figure 4.17 shows the electric-field intensity and current density in a lossless quad-rat race having a simpler layout than the actual design. In practice, the electromagnetic (EM) design of a quad-rat-race (and also an ordinary rat race) involves taking account of reflections at the T-junctions, losses, and the imaginary parts of characteristic impedances of the transmission lines, not usually considered in ideal circuit theoretic treatment. In our final design, the access lines and the ring are implemented using the same type of microstrip transmission line having a characteristic impedance close to $50\,\Omega$. Its signal strip width is $9\,\mu m$. We did not use a $35\text{-}\Omega$ line for the ring because only the $50\text{-}\Omega$ line was experimentally characterized and modeled for circuit simulation. The model was used to predict the characteristics of the quad-rat-race before the actual layout was done. EM-simulated characteristics of the actual design are shown in Figure 4.18. The somewhat high reflection as seen from $50\text{-}\Omega$-referenced ports can be attributed to the use of the $50\text{-}\Omega$ line for the ring. We designed matching networks taking that into consideration.

4.1.2.4   Overall TX design

Figure 4.19 shows the overall schematic of the 300-GHz CMOS TX. It receives a modulated IF signal at IF_1 and a W-band LO signal. The modulated signal is upconverted to IF_2 (around $100\,GHz$) by a double-balanced mixer. At this point, we must superpose IF_2 and LO so that we can feed the cubic mixers with ($IF_2 + LO$). However, a large area would be required if we used another passive device to superpose IF_2 and LO. Instead, we designed the double-balanced mixer such that the LO and $\overline{LO}$ ports are very close to the output ports. The intentional capacitive coupling thus introduced

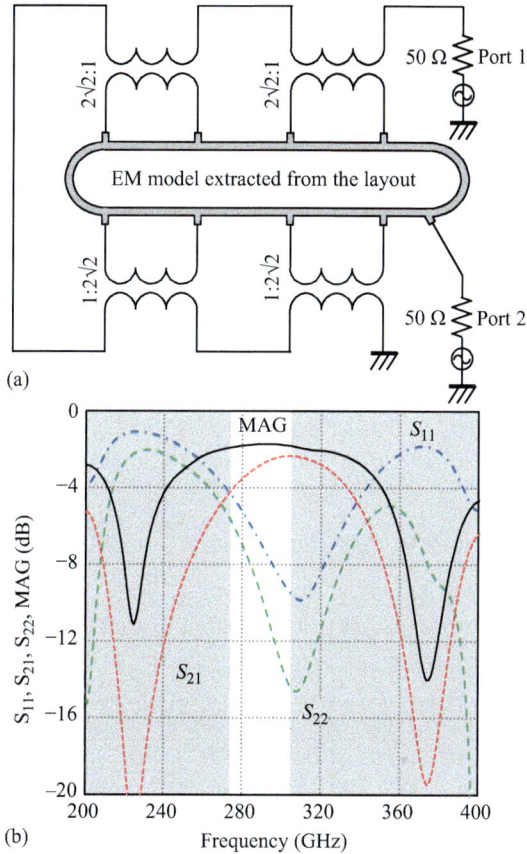

Figure 4.18   *(a) Simulation setup for bundling together the input ports of a quad-rat-race and (b) simulated frequency response of the quad-rat-race (actual design). Copyright 2016 IEEE. Reproduced, with permission, from [1]*

provides an LO leakage path and results in the desired superposition of IF_2 and LO at the output of the mixer.

After 32-way power splitting (Figure 4.10), the signal is upconverted to 300 GHz by 16 cubic mixers (Figure 4.7). The power combiner consists of four quad-rat-races and transmission lines (Figure 4.14). Figure 4.20 shows a die micrograph of the TX. The die size is 2 mm × 3 mm.

4.1.3   Analysis of cubic mixer [1]

The cubic mixer generates unused spectral components, as is clear from (4.1). Currently, no particular care is exercised to filter out these unused components. It is thus natural that they will adversely affect the desired signal and degrade the TX performance. In this section, we examine the frequency-dependent signal-to-noise ratio (SNR) of the cubic mixer.

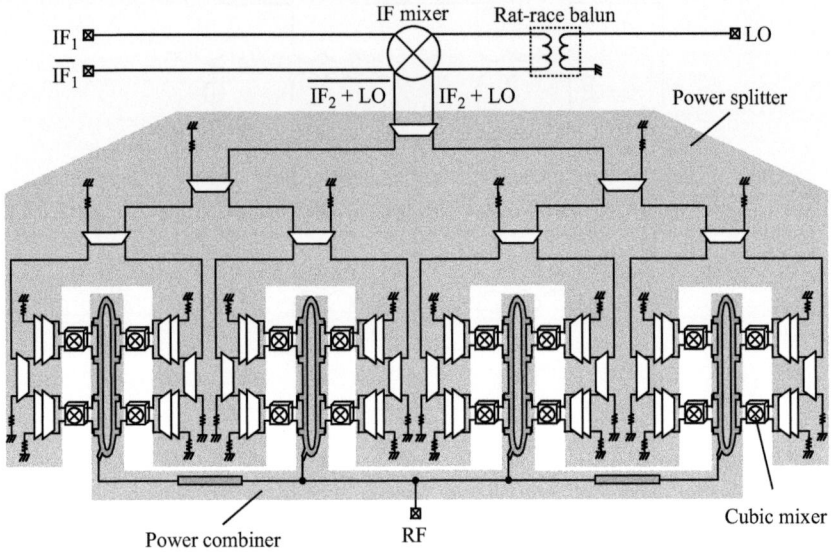

Figure 4.19   Schematic diagram of the 300-GHz CMOS transmitter. Copyright 2016 IEEE. Reproduced, with permission, from [1]

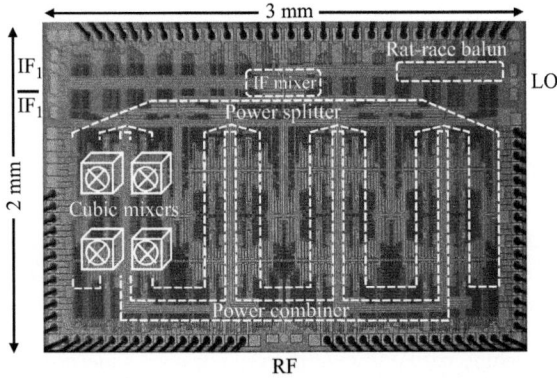

Figure 4.20   Die micrograph of the 300-GHz CMOS transmitter. Copyright 2016 IEEE. Reproduced, with permission, from [1]

Let the frequency separation between IF_2 and LO be Δf,

$$\Delta f = |LO - IF_2|. \tag{4.5}$$

In the present implementation, the output from the first upconversion mixer (Figure 4.19) contains not only IF_2 and LO but also the image of IF_2 at IF_{2IM}. If $IF_2 < LO$, as in Figure 4.21(a),

$$IF_{2IM} = IF_2 + 2\Delta f. \tag{4.6}$$

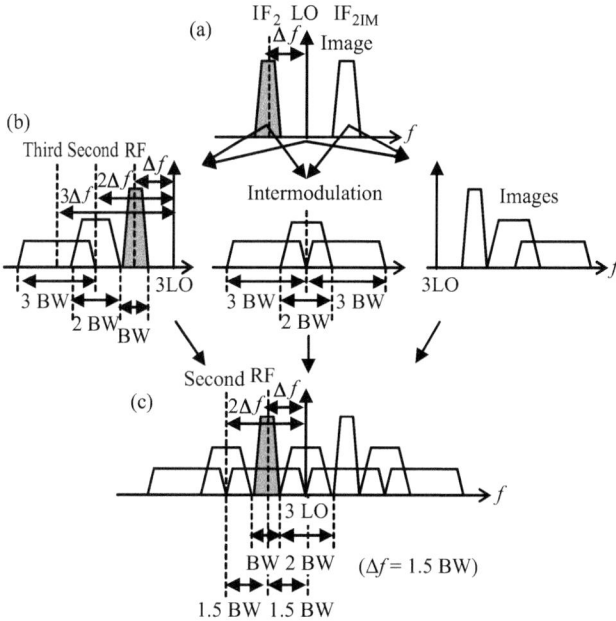

Figure 4.21   *(a) Input power spectrum fed into the cubic mixer. $IF_2 < LO < IF_{2IM}$ in this example. (b) Spectral components generated by the cubic mixer. (c) Output power spectrum of the cubic mixer obtained by superposing all the components shown in (b). Copyright 2016 IEEE. Reproduced, with permission, from [1]*

Thus, the cubic mixer generates the spectral components shown in Figure 4.21(b) [22], including the terms in (4.1) from IF_2 and IF_{2IM} [Figure 4.21(b), left and right], and the intermodulation products from IF_2 and IF_{2IM} [Figure 4.21(b), center].

If Δf is too small, the desired signal, RF, may interfere with the second-order term, $3LO \cdot IF_2^2$, from (4.1). This can be avoided by making $\Delta f \gtrsim 1.5BW$, where BW is the bandwidth of RF. Figure 4.22 shows system simulation results with three different values of Δf. The SNR of the cubic mixer degrades if $\Delta f \lesssim 1.5BW$. Of course, Δf cannot be increased indefinitely, and its upper bound is dictated by the overall TX bandwidth (Figure 4.22, right).

Another more fundamental noise source is the lower side, third-order intermodulation product [Figure 4.21(b), center]. It always interferes with RF itself, regardless of the value of Δf, as shown in Figure 4.21(c). In addition, as is clear from Figure 4.22, other spurious components interfere with the neighboring channels. The present configuration, therefore, does not allow multiple channels to be used simultaneously. A fundamental solution would be to design a TX such that IF_{2IM} does not exist in the input to the cubic mixer. This can be achieved, for example, by using a quadrature modulator. Another less complete but easier fix applicable to this TX is to choose LO and IF_2 in such a way that IF_{2IM} is driven outside the bandwidth of the

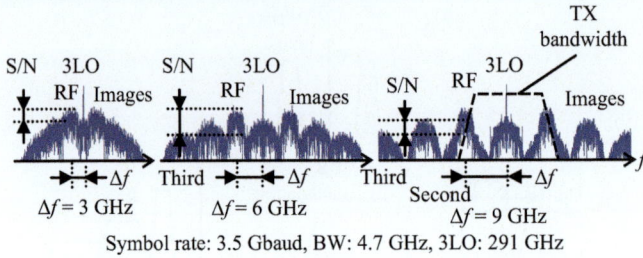

S/N 3LO   S/N   3LO   3LO   TX
RF Images   RF   Images   S/N   RF   bandwidth
                                                    Images

→⊢←Δf   Third →⊢←Δf   Third →⊢   ⊣←Δf
Δf = 3 GHz   Δf = 6 GHz   Second
                                    Δf = 9 GHz
Symbol rate: 3.5 Gbaud, BW: 4.7 GHz, 3LO: 291 GHz

*Figure 4.22   Simulated output spectrum of the cubic mixer for $\Delta f = 3$, 6, and
9 GHz. A root-raised-cosine filter with a roll-off factor of 0.35 was
applied to the input signal. "S/N" in the figure indicates that the
height is related to the SNR, but it has no quantitative meaning. The
result for $\Delta f = 6$ GHz gives the highest SNR. Copyright 2016 IEEE.
Reproduced, with permission, from [1]*

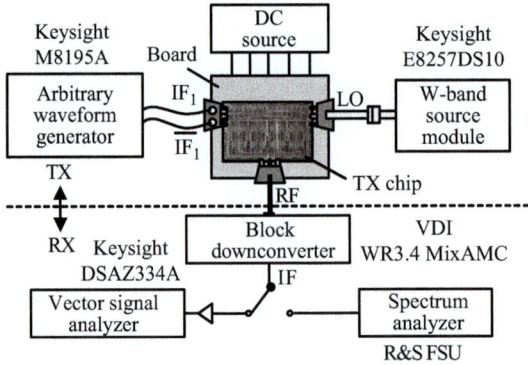

*Figure 4.23   Measurement setup. Copyright 2016 IEEE. Reproduced, with
permission, from [1]*

IF amplifiers for IF_2 in the power-splitting network (Figure 4.5) [23]. This will allow
higher data rates to be achieved over all six channels [23].

4.1.4   Transmitter performance [1]

Figure 4.23 shows the measurement setup. The arbitrary waveform generator generates a digitally modulated IF signal at IF_1 and feeds the TX chip differentially. The
W-band source module supplies the LO signal. The output RF signal at 300 GHz
is led to a block downconverter through a WR3.4 waveguide (WG) probe. The
downconverted signal is fed either to a vector signal analyzer or to a spectrum analyzer.

Figure 4.24 shows snapshots of the TX and the measurement system. The chip is
mounted on a printed circuit board (PCB). DC power is supplied through bond wires,
and high-frequency signals are supplied or measured via RF probes. Figure 4.25
shows the frequency up/down-conversion taking place in the TX and the measurement

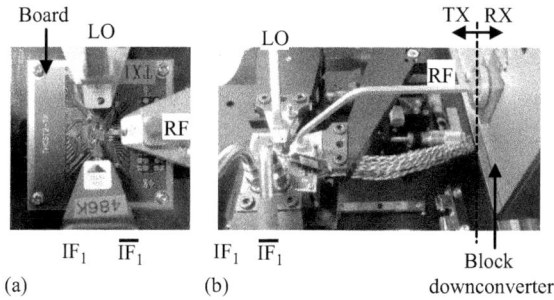

Figure 4.24 (a) TX printed circuit board (PCB) and (b) waveguide connection to the block downconverter. Copyright 2016 IEEE. Reproduced, with permission, from [1]

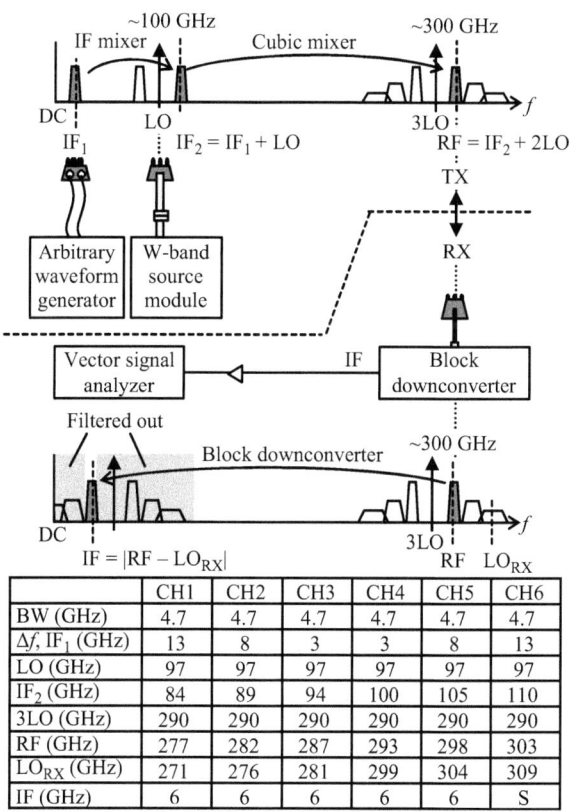

	CH1	CH2	CH3	CH4	CH5	CH6
BW (GHz)	4.7	4.7	4.7	4.7	4.7	4.7
Δf, IF_1 (GHz)	13	8	3	3	8	13
LO (GHz)	97	97	97	97	97	97
IF_2 (GHz)	84	89	94	100	105	110
3LO (GHz)	290	290	290	290	290	290
RF (GHz)	277	282	287	293	298	303
LO_{RX} (GHz)	271	276	281	299	304	309
IF (GHz)	6	6	6	6	6	S

Figure 4.25 Frequency upconversion in the transmitter and downconversion in the measurement system. Copyright 2016 IEEE. Reproduced, with permission, from [1]

LO = 96.6 GHz, 8 dB m, IF$_1$ = 9.8 GHz

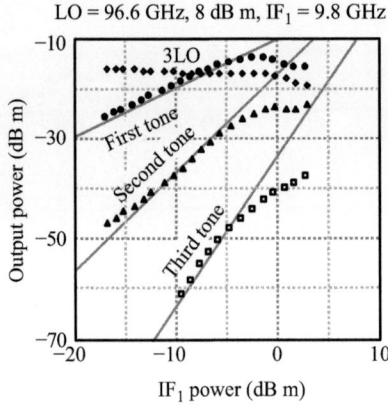

Figure 4.26   Output power as a function of IF$_1$ power. Copyright 2016 IEEE. Reproduced, with permission, from [1]

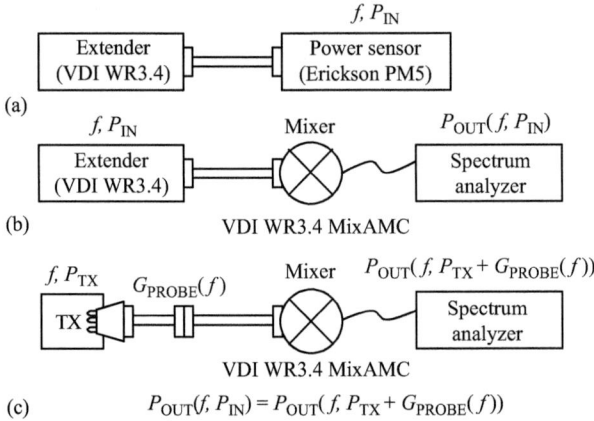

Figure 4.27   Calibration procedure for the output power measurement. (a) The output power, P_{IN}, of a WR3.4-band frequency extender is measured as a function of frequency f. (b) The frequency-dependent conversion loss of the block downconverter (mixer) is measured and tabulated as $P_{OUT}(f, P_{IN})$. (c) The TX output power P_{TX} (in dB m) is found from $P_{OUT}(f, P_{TX} + G_{PROBE}(f))$. The probe loss, $G_{PROBE}(f)$ (in dB), is characterized by another measurement. Copyright 2016 IEEE. Reproduced, with permission, from [1]

system. The first IF, IF$_1$, fed to the chip is upconverted to the second IF, IF$_2$, at around 100 GHz. The values of IF$_1$ and IF$_2$ are given in Figure 4.25. Then the signal is upconverted by the cubic mixer to about 300 GHz. The 300-GHz RF signal is down converted to 6 GHz by the block downconverter. Unwanted spectral components that reach the vector signal analyzer are filtered out before demodulation.

The measured output RF power (Figure 4.26), calibrated by the procedure shown in Figure 4.27, clearly shows linear dependence on the input IF$_1$ power as intended.

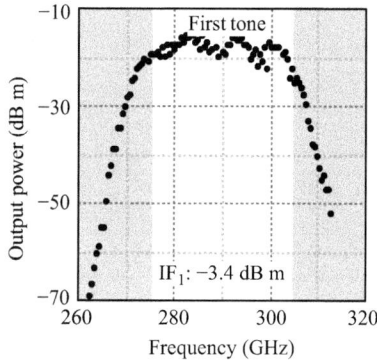

Figure 4.28   *Output RF power as a function of frequency. Copyright 2016 IEEE. Reproduced, with permission, from [1]*

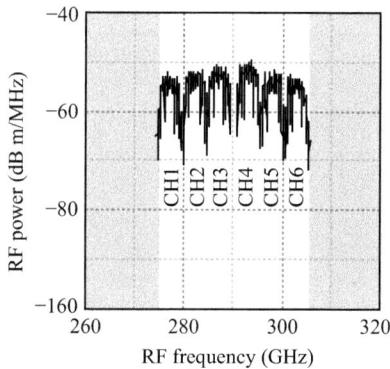

Figure 4.29   *Channel power spectra of a 3.5-Gbaud 32QAM signal for the six channels (Figure 4.2). Copyright 2016 IEEE. Reproduced, with permission, from [1]*

This linearity is crucial to successful QAM signal transmission. The peak output power was −14.5 dB m. The frequency dependence of the output power shown in Figure 4.28 indicates that RF transmission is possible for all six channels (Figure 4.2). Figure 4.29 shows the output power spectra of a 3.5-Gbaud 32QAM signal for the six channels. Each channel spectrum was measured separately with the spectrum analyzer [1]. The corresponding 32QAM signal constellations are shown in Figure 4.30. All channels achieved 17.5 Gbit/s with 32QAM. The gross aggregate data rate reached 17.5 × 6 = 105 Gbit/s. The TX consumes 1.4 W of DC power.

Figure 4.30 shows that the SNR depends strongly on the channel. This is because of the frequency-dependent characteristics of the cubic mixer, discussed in Section 4.1.3. Figure 4.31 shows the measured error-vector magnitude (EVM) and estimated SNR of a 3.5-Gbaud QPSK signal as a function of the center frequency of RF. The SNR was calculated from the following rough approximate formula [24]:

$$\mathrm{SNR} \simeq \frac{1}{\mathrm{EVM}^2},$$

(4.7)

Channel RF freq. (GHz)	CH1 275–279	CH2 280–284	CH3 285–289
Constellation (equalized)			
EVM	8.9%rms	4.8%rms	7.0%rms
Data rate	17.5 Gb/s	17.5 Gb/s	17.5 Gb/s
Channel RF freq. (GHz)	CH4 291–295	CH5 296–300	CH6 301–305
Constellation (equalized)			
EVM	7.1%rms	6.4%rms	5.9%rms
Data rate	17.5 Gb/s	17.5 Gb/s	17.5 Gb/s

Figure 4.30   32QAM signal constellations and error-vector magnitudes (EVMs) at 17.5 Gbit/s. Copyright 2016 IEEE. Reproduced, with permission, from [1]

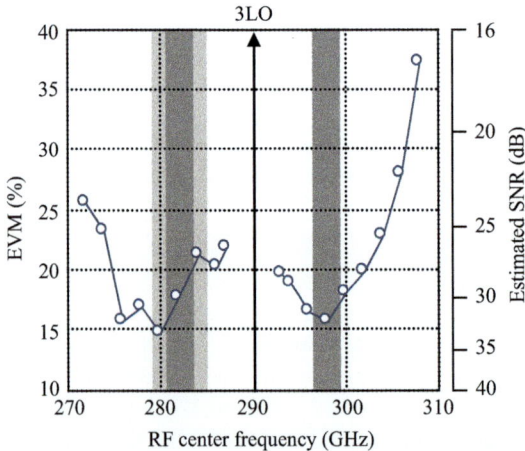

Figure 4.31   Measured EVM of a 3.5-Gbaud QPSK signal and estimated SNR plotted against the center frequency of RF. The shaded regions have a high SNR. Copyright 2016 IEEE. Reproduced, with permission, from [1]

or

$$\text{SNR (dB)} \simeq -20 \log_{10} \frac{\text{EVM (\%)}}{100}. \tag{4.8}$$

The measurement was made with a fixed LO ($3LO = 290\,\text{GHz}$), and $\Delta f = |3LO - RF|$ was swept. The result is consistent with the discussion in Section 4.1.3.

Table 4.1 Performance comparison of terahertz transmitters

Ref.	Technology	Freq. (GHz)	Mod.	Data rate (Gbit/s)	P_{out} (dB m)	P_{DC} (W)
[5]	32-nm SOI CMOS	210	OOK	20	4.6	0.24
[6]	250-nm InP DHBT	298	NA	NA	−2.3	0.45
[7]	130-nm SiGe BiCMOS	240	64QAM	1.02	7	0.54
[26]	35-nm GaAs mHEMT	240	8PSK	96	−3.5	NA
[27]	35-nm GaAs mHEMT	240	QPSK	64	−3.6	NA
[28]	130-nm SiGe HBT	314	NA	NA	−8	0.13
[10]	130-nm SiGe BiCMOS	434	ASK	10	−18.5	0.12
[11]	65-nm CMOS	260	OOK	NA	5 (EIRP)	0.69
[2]	SBD mixer	300	64QAM	0.032	−15	NA
[3]	SBD mixer	340	16QAM	3	−17.5	NA
[8]	130-nm InP HBT	630	NA	NA	−30	0.65
[9]	250-nm InP HBT	300	QPSK	50	NA	NA
[12]	65-nm CMOS	240	QPSK	16	0	0.22
This work	40-nm CMOS	275–305 282	32QAM	17.5×6 30	−14.5	1.4

EIRP, equivalent isotropically radiated power; SBD, Schottky barrier diode.

SNR degrades when Δf is either too small or too large. Figure 4.31 explains why CH2 gave the lowest EVM in Figure 4.30. Also, the per-channel data rate of 17.5 Gbit/s is limited by the noisiest channel. Higher per-channel data rates should therefore be achievable in the shaded regions in Figure 4.31. With 3LO = 290 GHz and RF = 282 GHz (Δf = 8 GHz), we achieved 30-Gbit/s signal transmission with 32QAM.

Table 4.1 shows a summary of performances of state-of-the-art terahertz TXs. GaAs and InP technologies provide a very high f_{max} and hence high performance. Conventional CMOS TXs adopt a frequency-multiplier-last architecture [11,12]. The strength of this 300-GHz TX is that it supports QAM despite the fact that it operates above the transistor f_{max}. Note that the data rates quoted in Table 4.1 do not always represent the highest value achievable by the respective TX. This is because some of the experimental data are from wireless transmission experiments and some are from direct ("wired") measurements. It is also possible that the measured performance data were limited by the availability or performance of the test equipment. All the measurement data presented here are from wired measurements. The results of wireless measurements with the improved setting [23] mentioned at the end of Section 4.1.3 are presented elsewhere [25].

4.1.5 Doubler-based transmitter [40]

A CMOS transmitter operating above f_{max} achieved 28 Gbit/s at 300 GHz [1,25]. Although none of these TXs can generate high enough output power for line-of-sight links of tens of kilometers, perhaps driving a Watt-level traveling-wave tube, reportedly under development, might be what is required of IC-based terahertz TXs, if they are to be used for such applications. Our TX also operates above f_{max} and therefore massive power combining is required to make up for the lack of PAs. The CMOS TX in [1] adopted a tripler-based upconversion technique suitable for many-way power combining. However, its output power spectrum contained spurious components (tripled LO and unwanted mixing products including images) that would cause interference. There are known techniques for suppressing such unwanted components, including the Hartley architecture shown in Figure 4.32(a). However, most such techniques typically require multiple double-balanced mixers and will face layout difficulties when power combining is attempted.

4.1.5.1   Architecture

Figure 4.32(b) shows the architecture adopted in this work. The layout of its power-splitting and -combining networks has no crossovers. Upconversion is performed twice ($f_{IF1} \approx 10\,GHz \rightarrow f_{IF2} \approx 155\,GHz \rightarrow f_{RF} \approx 300\,GHz$) in this circuit and the second upconversion is done using a doubler-based mixer, which, following [1], could be called a "square (or quadratic) mixer." Since quadratic nonlinearity of a MOSFET is stronger than its cubic counterpart, higher output power can be expected than from a tripler-based "cubic mixer" [1]. If the signal fed to a square mixer consists of two spectral components, LO (at $f_{LO} \approx 145\,GHz$) and IF_2, the mixing product

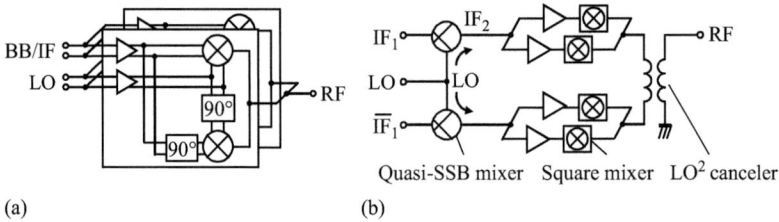

(a)   (b)

Figure 4.32   *Single-sideband (SSB) mixers. Possible 300-GHz-band CMOS transmitter architectures using (a) Hartley-type mixers and (b) quasi-SSB mixers. Copyright 2017 IEEE. Reproduced, with permission, from [40]*

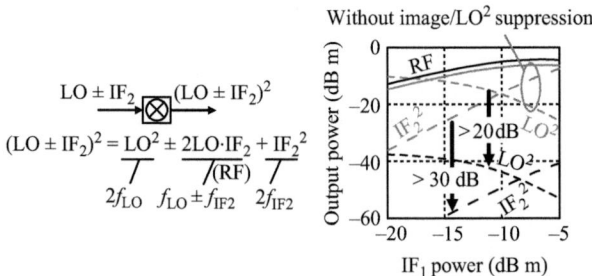

Figure 4.33   *The schematic concept and simulated input–output characteristics of a square mixer. The RF signal linearly up-converted by the input (IF_1). Unnecessary signals generated by a square mixer can be suppressed. Copyright 2017 IEEE. Reproduced, with permission, from [40]*

$RF = 2LO \cdot IF_2$ shows linear dependence on the input power as shown in Figure 4.33. This linearity is essential for supporting QAM. A schematic of the square mixer is shown in Figure 4.34(a). Figure 4.34(b) shows a comparison of the large-signal characteristics of a square mixer and cubic mixer. When the IF_2 and LO input powers are both 5 dB m, the RF output power of the square mixer is 6.6 dB larger than that of the cubic mixer.

4.1.5.2 Suppression of unwanted signals

If the input additionally contains IF_2's image IF_{2IM} (at $f_{IF2IM} \approx 135\,GHz$), it is also upconverted. To prevent that without complicating the layout, we perform the first upconversion using what we call a "quasi-single-sideband (SSB) mixer." Figure 4.35 explains the operation of the upconversion circuit in Figure 4.32(b) in more detail. The quasi-SSB mixer sits at the input in Figure 4.35. Schematically, it looks like a double-sideband (DSB) mixer. For a given input IF signal frequency f_{IF1} and a mixer output frequency f_{out}, we can think of two distinct cross-frequency response functions, $K_U(f_{out}, f_{IF1}, f_{LOU})$ and $K_L(f_{out}, f_{IF1}, f_{LOL})$, for upper sideband (USB) and lower sideband (LSB) generation, respectively (Figure 4.36(a), where $f_{LOU} = f_{out} - f_{IF1}$ and $f_{LOL} = f_{out} + f_{IF1}$ are the respective LO frequencies. Quasi-SSB mixer's matching

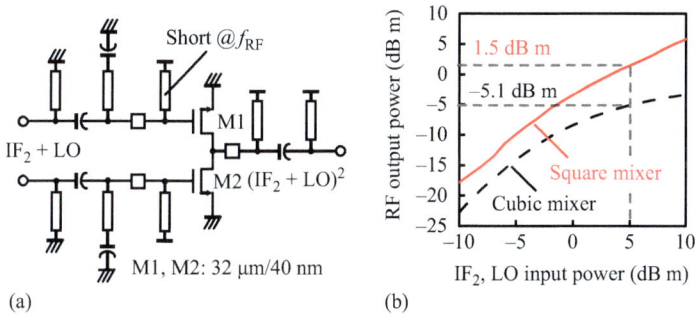

Figure 4.34 (a) Schematic of a square mixer and (b) characteristics comparison between a cubic mixer and a square mixer. Copyright 2017 IEEE. Reproduced, with permission, from [40]

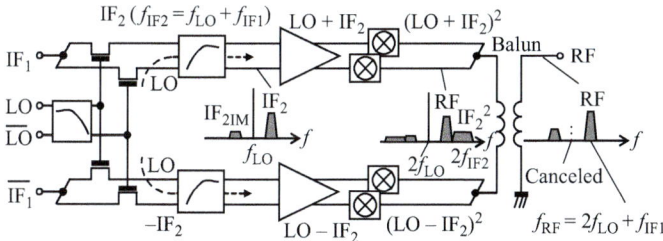

Figure 4.35 A block diagram of a 300-GHz-band CMOS transmitter using quasi-SSB mixers and square mixers. Unnecessary signals generated by the square mixers are suppressed by using two paths. Copyright 2017 IEEE. Reproduced, with permission, from [40]

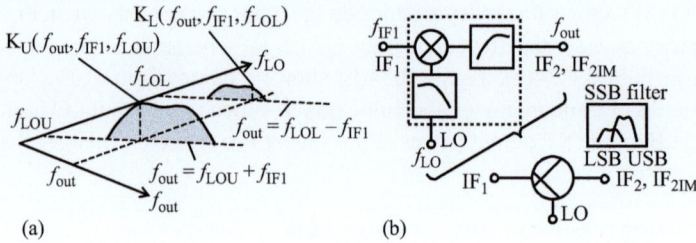

Figure 4.36   *Operation of a quasi-SSB mixer: (a) conceptual frequency response of a quasi-SSB mixer and (b) configuration of a quasi-SSB mixer. Copyright 2017 IEEE. Reproduced, with permission, from [40]*

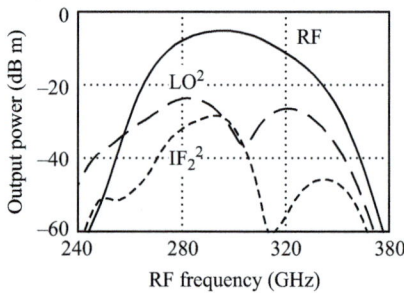

Figure 4.37   *Simulated frequency response of a square mixer. Copyright 2017 IEEE. Reproduced, with permission, from [40]*

networks, shown symbolically as filters in Figure 4.36(b), are designed such that $|K_U| > |K_L|$. Such asymmetric design for USB and LSB is possible if f_{IF1} is very high (≥ 10 GHz) as in our terahertz TX. Thus, the quasi-SSB mixer upconverts IF_1 primarily into IF_2 (the USB) as shown in the inset Figure 4.35. Since the quasi-SSB mixer is single-balanced and the MOSFETs constituting it are biased, LO necessarily "leaks out." The resulting $(LO + IF_2)$ at the mixer output is squared by a square mixer shown in Figure 4.34(a). Finally, the output balun subtracts $(LO - IF_2)^2$ from $(LO + IF_2)^2$ and cancels out LO^2 and IF_2^2, as shown in the lower right graph in Figures 4.34(b) and 4.37.

4.1.5.3   Double-rat-race

Power splitting shown in Figure 4.32(b) is done using a planar passive device called a "double-rat-race" (Figure 4.38(a)), which is similar to the quad-rat-race introduced in [1]. The former can perform a four-way power splitting. The access lines and the ring of an ordinary rat race (Figure 4.38(b)) are designed to have characteristic impedances of Z_0 and $\sqrt{2Z_0}$, respectively, whereas the characteristic impedance of the ring of a double-rat-race should be Z_0. To see this, suppose that a double-rat-race power splitter (Figure 4.38(a)) is excited from its input port and that all the output ports are terminated with Z_0. Then, node A becomes virtual ground because clockwise

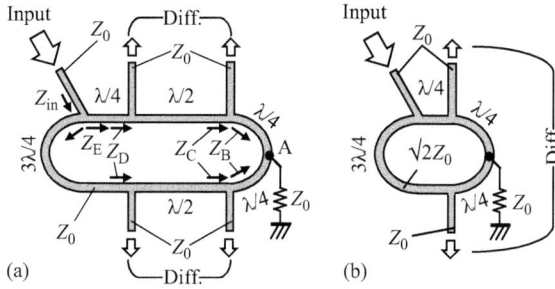

Figure 4.38   (a) A double rat race for power splitting and (b) an ordinary rat-race balun. Copyright 2017 IEEE. Reproduced, with permission, from [40]

Table 4.2   Performance comparison

	[9]	[26]	[29]	[7]	[1]	This work	
Technology	250 nm InP	35 nm GaAs	35 nm GaAs	0.13 μm SiGe	40 nm CMOS	40 nm CMOS	
Freq. (GHz)	300	240	300	240	300	302	289–311
Modulation	QPSK	8PSK	QPSK	64QAM	16QAM	32QAM	128QAM
P_{out} (dB m)	–	−3.5	−4	7	−14.5	−5.5	
P_{DC} (W)	–	–	–	0.54	1.4	1.4	
Data rate (Gbit/s)	50	96	64	1.02	28	105	24.64 × 6

and counterclockwise traveling waves emanating from the input T-junction reach there 180° out-of-phase. The $\lambda/4$-transformers connected to node A give $Z_B = \infty$. Thus, $Z_C = Z_0 \| Z_B = Z_0$. Moving further back toward the input port, the $\lambda/2$-sections perform no impedance transformation, and therefore $Z_D = Z_0 \| Z_C = Z_0/2$. The final $\lambda/4$-transformers leading to the input T-junction should transform Z_D into $Z_E = 2Z_0$ so that $Z_{in} = Z_E \| Z_E = Z_0$. The characteristic impedance of the ring must therefore be Z_0. Matching at the other ports can be analyzed likewise. This low-mismatch-loss design contributes to higher output power.

4.1.5.4   Measurement

Figure 4.39 shows an overall TX schematic. The LO signal comes from an on-chip tripler driven by an external signal $LO^{1/3}$. Modulated IF_1 is supplied differentially from an arbitrary waveform generator and is upconverted to IF_2 by a pair of quasi-SSB mixers. Four double-rat-races perform power splitting. 16-way power combining at 300 GHz is done using 2 H trees and a rat-race balun. Figure 4.40 shows the measurement setup. IF_1, $LO^{1/3}$, and RF are supplied/measured through RF probes shown in Figure 4.41. Figure 4.42(a) shows that RF power depends linearly on IF_1 power and that higher order mixing products have much lower power. Measured USB and LSB frequency responses in Figure 4.42(b) show that LSB suppression is in effect.

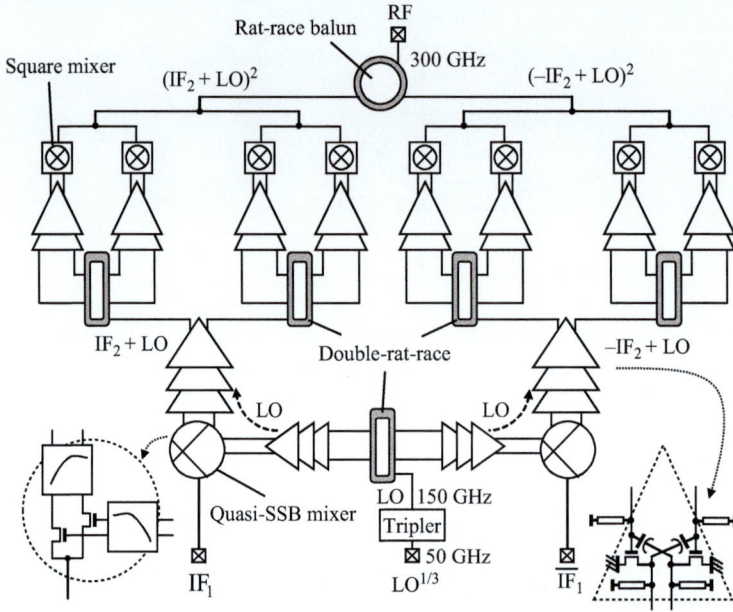

Figure 4.39   *Overall schematic of a 300-GHz-band CMOS transmitter. Copyright 2017 IEEE. Reproduced, with permission, from [40]*

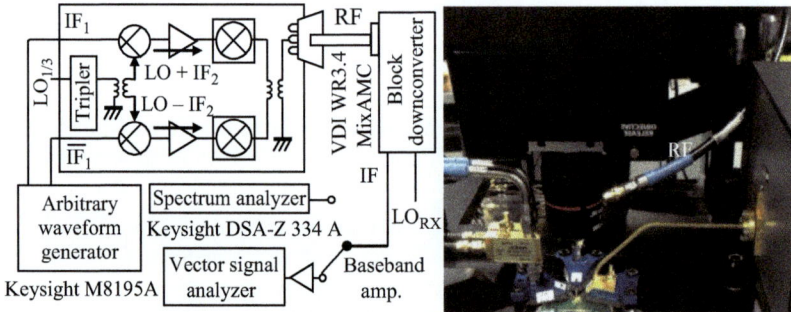

Figure 4.40   *Measurement setup. Copyright 2017 IEEE. Reproduced, with permission, from [40]*

The TX achieves its highest single-channel data rate of 105 Gbit/s with 32QAM. Alternatively, the TX can do 24.64 Gbit/s/ch over 6 128QAM channels with a fixed f_{IF1} and six different values of f_{LO} and then the aggregate data rate reaches 147 Gbit/s. The corresponding signal constellations and output power spectra are shown in Figure 4.43. Note that the image of CH6, $CH6_{IM}$, has the same frequency as CH1, yet CH1 is not suppressed like $CH6_{IM}$. This clearly indicates that the quasi-SSB mixer is different from a DSB mixer followed by a low-pass filter, which would have suppressed $CH6_{IM}$ and CH1 equally. Although suppression of the images is not satisfactory, the measurement results prove the concept of the quasi-SSB mixer. LO_2 is

TX PCB

Figure 4.41 *Printed circuit board for the transmitter measurement. RF probes are also shown. Copyright 2017 IEEE. Reproduced, with permission, from [40]*

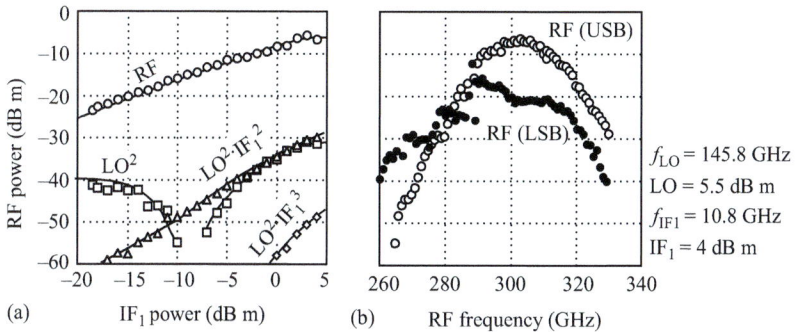

Figure 4.42 *Measured characteristics of a 300-GHz-band CMOS transmitter: (a) input–output responses of wanted and unwanted signals and (b) frequency responses of upper and lower sidebands. Copyright 2017 IEEE. Reproduced, with permission, from [40]*

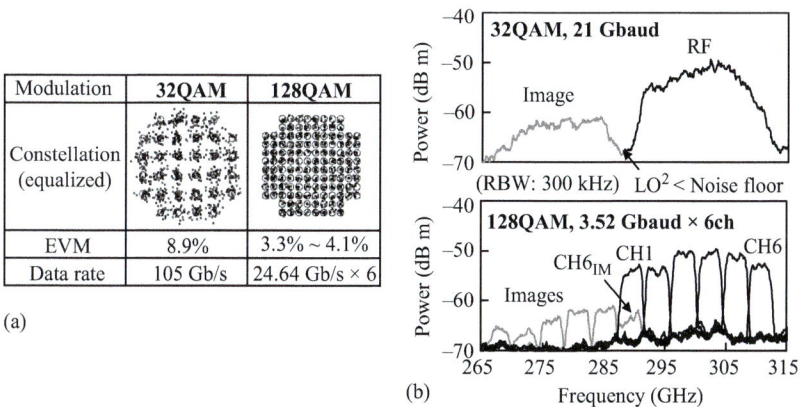

Modulation	32QAM	128QAM
Constellation (equalized)		
EVM	8.9%	3.3% ~ 4.1%
Data rate	105 Gb/s	24.64 Gb/s × 6

Figure 4.43 *Measured (a) constellations and (b) spectra. Copyright 2017 IEEE. Reproduced, with permission, from [40]*

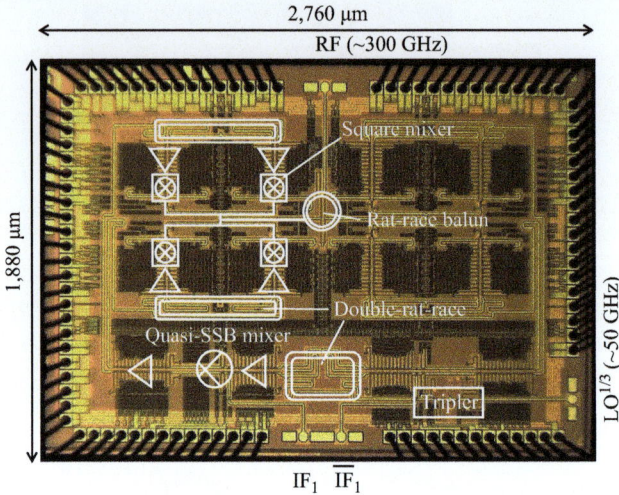

Figure 4.44   Chip micrograph. Copyright 2017 IEEE. Reproduced, with permission, from [40]

satisfactorily canceled out and not visible in the measured spectra. Table 4.2 shows performance comparison of recent IC-based terahertz TXs. The peak output power of our TX is $-5.5\,dB\,m$ and the power consumption is $1.4\,W$. A die micrograph of the TX is shown in Figure 4.44. The die area is $2.76 \times 1.88\,mm^2$.

4.1.6   Transmitter module [51]

Terahertz (THz) wireless technology could bring us ultrahigh-speed wireless communication systems because of the vast available frequency band it offers [30]. We recently reported a 300-GHz CMOS TX that realized wireless digital data transmission at 56 Gbit/s over 5 cm [31]. The CMOS TX chip was not packaged in that experiment and the transmitting horn antenna was fed via a WR3.4 WG probe. If such a TX is to be put to practical use, the chip must be packaged and the TX be modularized. To have an antenna gain of 20 dB i or higher [32], on-chip and on-board antennas are currently not an option. To connect a CMOS chip to a high-gain antenna, a CMOS-chip-to-WG transition is required. However, it is difficult to realize a direct transition of a CMOS chip to a WG because precise cutting and back-grinding of a chip is required. To avoid that difficulty, a CMOS-chip-to-WG transition should consist of two parts. One is a connection of a transmission line on a CMOS chip onto a board, and another is a transition of a transmission line on the board to a WG. A transmission-line-to-WG needs a back-short structure for good impedance matching. Recently, a microstrip-to-WG transition integrated in a multilayer low-temperature co-fired ceramics (LTCC) substrate at 300 GHz was reported [33]. It realized a vertical hollow WG with a back-short structure inside the multilayer board. Although LTCC has small loss tangent [33], it is costly and not very suitable for mass production. In terms of the cost, glass epoxy PCBs are superior to LTCC boards. Although the loss tangent of glass epoxy is several times that of LTCC, it seems possible to utilize a multilayered glass epoxy PCB to build a WG transition with acceptable losses.

In this subsection, we present a CMOS-chip-to-WG transition integrated into a low-cost multilayered glass epoxy PCB. In addition, we show the wireless performance of the 300-GHz CMOS TX module.

4.1.6.1 Structure of the 300-GHz CMOS transmitter module

Figure 4.45(a) shows the layers of the glass epoxy PCB. The CMOS TX chip is mounted near the center of the PCB by gold stud bump flip-chip bonding underneath a WR3.4 WG flange. The latter has a bore on the bottom side for accommodating the

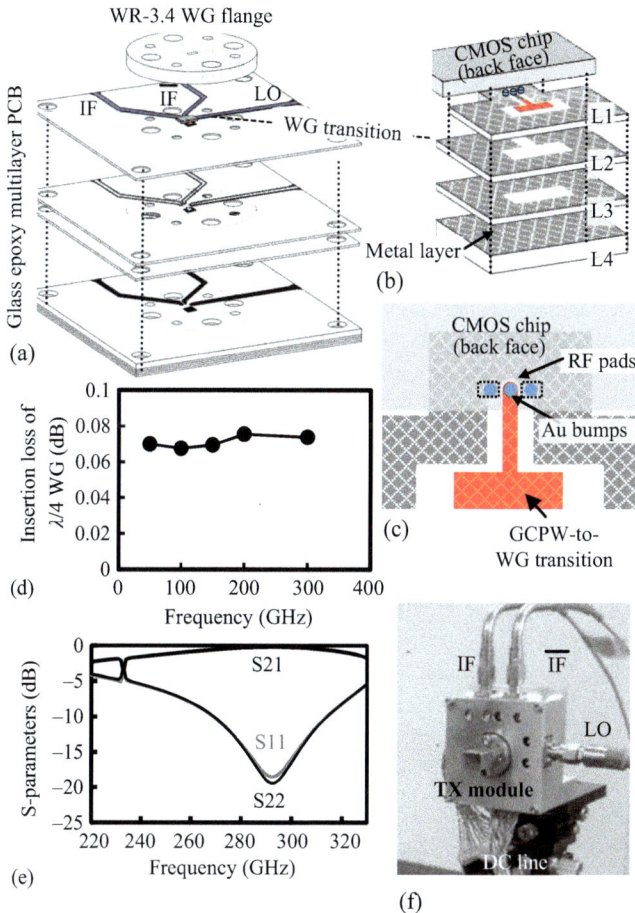

Figure 4.45 Structure of (a) the module board and (b), (c) waveguide transition. (d) The insertion loss of a $\lambda/4$-long waveguide as a function of frequency. (e) The simulated S-parameters of the waveguide transition. (f) Photo of the transmitter module with a horn antenna. Copyright 2018 IEEE. Reproduced, with permission, from [51]

chip. The IF and LO signals are supplied from external sources via coaxial connectors (not shown) mounted on the edges of the PCB and led to the chip through grounded coplanar WGs (GCPWs). Figure 4.45(b) shows a magnified view of the GCPW-to-rectangular-WG transition. The pads for the 300-GHz RF signal are 75-μm-pitch ground-signal-ground (GSG) pads. The height of the bumps are approximately 20 μm. The GCPW-to-WG transition has a vertical hollow cut into the four-metal-layer PCB, serving as a back short. The depth of the back-short structure should ideally be λ/4, where λ is the wavelength of the RF signal. Since the nominal relative permittivity of the glass epoxy board is 3.26 at 10 GHz and the multilayered structure restricts the realizable depths to discreate values, the depth of 135 μm is selected. Figure 4.45(c) shows the insertion loss of a λ/4-long WG as a function of frequency. The nominal loss tangent of the glass epoxy is 0.0076 at 10 GHz. ANSYS HFSS was used for the simulation. The insertion loss of the λ/4-long WG at 300 GHz was almost the same as that at lower frequencies. The dimensions of the rectangular probe (Figure 4.45(d)) made of the top metal layer is determined to give good impedance matching between the transmission line (GCPW) and the WR3.4 rectangular WG. In addition, the dimensions of the vertical hollow WG are adjusted for impedance matching. The dimensions are 0.8 times the inner dimensions of a WR3.4 WG. Figure 4.45(e) shows S parameters of the GCPW-to-WG transition simulated before fabrication. The port 1 reference plane is at the center of the GSG pads on the PCB. The port 2 reference plane is on the top side of the WG flange. Permittivity and loss tangent values at 10 GHz were used in the simulation for the 300-GHz band because the values at these frequencies were unknown. As shown in Figure 4.45(e), S_{21} is −0.23 dB, and S_{11} and S_{22} stay below −10 dB over a 45-GHz bandwidth. Figure 4.45(f) shows the TX module with a horn antenna attached directly to the WR3.4 interface.

4.1.6.2 Measurement results

The performance of the 300-GHz CMOS TX module was measured by using a vector network analyzer (VNA) with a WR3.4-band frequency extender. The TX module was directly connected to it. Figure 4.46 shows the measured and simulated reflection

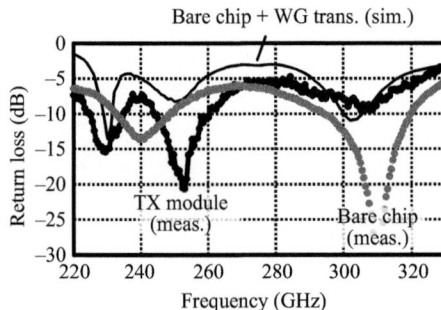

Figure 4.46 *Reflection coefficients of the RF port of the module and the bare chip. Solid line is the simulated result using a more precise simulation model identified after the fabrication. Copyright 2018 IEEE. Reproduced, with permission, from [51]*

coefficients of the RF port of the module. The accuracy of the post-fabrication simulation was improved by using a more precise simulation model and extracting the permittivity and the loss tangent of the board at 300 GHz. Moreover, an output impedance of the CMOS chip was measured and used in the simulation. The mismatch loss of the module at around 300 GHz was higher than that of the bare chip due to imprecise information available before fabrication. Figure 4.47(a) shows the measured output power as a function of the input IF power. The undesired signals (LO leak and IF22) were suppressed effectively. The peak output power of the desired signal was -13.5 dB m. It indicates that the packaging loss is approximately 8 dB because the output RF power of the CMOS chip is -5.5 dB m [31]. Figure 4.47(b) shows the frequency response of the output RF power of the module. The 3-dB bandwidth was approximately 10 GHz with a center frequency of 305 GHz.

Figure 4.48 shows a photo of wireless performance measurement setup. The IF and LO signals were supplied to the module by an arbitrary waveform generator and a signal generator, respectively. A pair of WR3.4 horn antennas with a gain of 24 dB i were used. A block-downconverter on a guide rail and an IF amplifier were used as

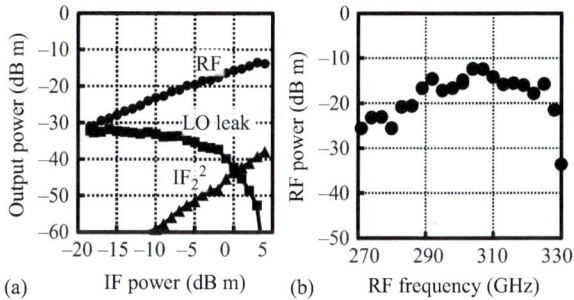

Figure 4.47   Measured output power of the transmitter module (a) as a function of input IF power and (b) as a function of frequency. LO power is 6 dB m. Copyright 2018 IEEE. Reproduced, with permission, from [51]

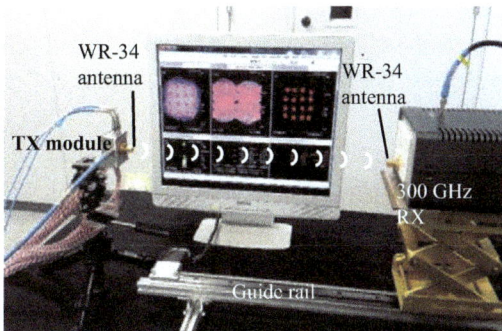

Figure 4.48   Measurement setup for wireless demonstration of the transmitter module. Copyright 2018 IEEE. Reproduced, with permission, from [51]

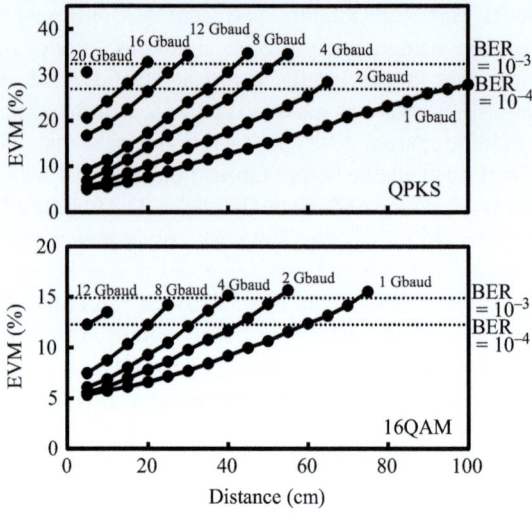

Figure 4.49   Measured error-vector magnitude as a function of distance. Copyright 2018 IEEE. Reproduced, with permission, from [51]

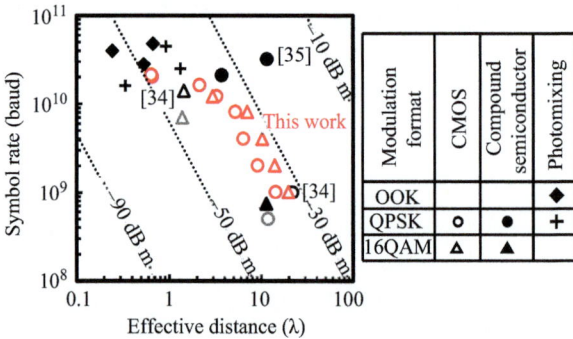

Figure 4.50   Comparison of symbol rate vs effective distance. Dotted lines show constant-FoM lines [32]. Copyright 2018 IEEE. Reproduced, with permission, from [51]

a receiver (RX) system. To evaluate the performance of the wireless link, the EVM of the received signal was measured by a vector signal analyzer while varying the modulation format, the symbol rate, and the antenna-to-antenna distance. A channel equalizer built into the demodulation software was applied. The modulation formats used in the experiment were QPSK and 16QAM.

The measurement results are shown in Figure 4.49. The highest data rate of 48 Gbit/s with 16QAM was achieved over a transmission distance of 5 cm. Wireless transmission over a distance exceeding 1 m could be possible at a data rate of 2 Gbit/s with QPSK. Figure 4.50 shows a comparison of symbol rates vs effective distances of terahertz wireless links including a figure of merit (FoM) [32]. The closer to the upper right corner of the graph, the higher the wireless performance. Our module has relatively high wireless performance.

4.1.6.3 Conclusion

We presented a 300-GHz CMOS TX module with a CMOS-chip-to-WG transition built in a low-cost multilayer glass epoxy PCB. The CMOS chip was flip-chip bonded onto the PCB. A GCPW-to-WG transition was realized by a vertical hollow WG structure cut in the PCB. The measured output power and 3-dB bandwidth of the module were approximately -13.5 dB m and 10 GHz, respectively. A wireless link with the module achieved a data rate of 48 Gbit/s with 16QAM. Although there is still room for improvement, the possibility of achieving high wireless performance using low-cost PCBs and standard flip-chip bonding even at 300 GHz is encouraging.

4.2 Receiver

If the highest $f_{\max}$ available from the process to be used is below the operating frequency as in our case, no low-noise amplifier (LNA) can be built, and an LNA-less RX architecture must be adopted. In this section, we present a 300-GHz RX in 40-nm CMOS, operating above $f_{\max}$. There do not appear to be as many architectural degrees of freedom that the designer can play with in THz RX design as compared to THz TX design [36]. Once a choice of process technology has been made, the remaining most important problem is how to make the downconversion mixer. The nonlinear devices available from the 40-nm CMOS process that we use are MOSFETs, pn diodes and Schottky diodes. Although Schottky diodes fabricated using CMOS technology have recently been actively studied [37,38] and are becoming attractive, our present take on the choice of nonlinear devices is that MOSFETs are, overall, still the best.

4.2.1 Doubler-last LO driver [45]

Figure 4.51 shows the schematic diagram of our 300-GHz RX. The circuit consists of two parts: the downconversion mixer and the LO multiplier chain. The mixer part is composed of a balun and a double-balanced mixer. The LO multiplier part consists of a frequency tripler, power splitters, driver amplifiers, and frequency doublers. An LO signal at around 50 GHz ($LO^{1/6}$) is supplied from an off-chip signal source and is upconverted to about 300 GHz by the multiplier chain. The downconversion mixer outputs an IF signal at several GHz. Details of the design are explained below.

4.2.1.1 300 GHz downconversion mixer

The downconversion mixer, shown in Figure 4.51, is a double-balanced fundamental passive transistor mixer consisting of four 32-μm-wide transistors. The LO and RF are applied to their gates and drains, respectively. Although a fundamental mixer requires a higher LO frequency than a harmonic mixer, a higher conversion gain and a lower noise figure (NF) can be expected [39]. Matching networks for the mixer core are designed to cover an LO frequency range from 280 to 320 GHz with an RF frequency at 300 GHz. The matching networks are implemented using shunt and open stubs, as shown in Figure 4.51. Its simulated conversion gain is shown in Figure 4.52(a). The performance of the mixer also depends on the power level of the LO signal and its flatness over the desired bandwidth.

4.2.1.2 LO multiplier chain

A $6\times$ multiplier chain is designed so that high LO power is obtained over a wide bandwidth. An external LO signal, $LO^{1/6}$, at around 50 GHz is first tripled, then split into differential signals by a rat-race balun, and then amplified by driver amplifiers (Figure 4.51). The resulting signal at around 150 GHz undergoes 8-way power splitting by a pair of double-rat-race power splitters [40]. The signal frequency is then doubled by the four differential doublers and the LO signal at around 300 GHz is obtained. Subsequent power combining is done using a transmission-line H tree. Figure 4.52(b) shows the simulated output power of the LO multiplier chain. The output power is -4.4 dB m and the input-referred 3-dB bandwidth (at $LO^{1/6}$) is 5.8 GHz. The output bandwidth at RF is 34.8 GHz, from 277.2 to 314.2 GHz.

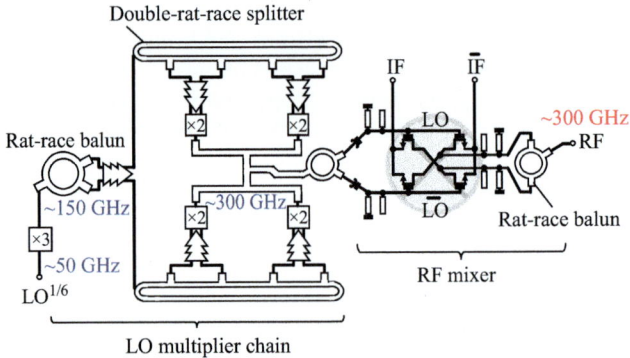

Figure 4.51   Schematic diagram of the 300-GHz receiver. Copyright 2017 IEEE. Reproduced, with permission, from [45]

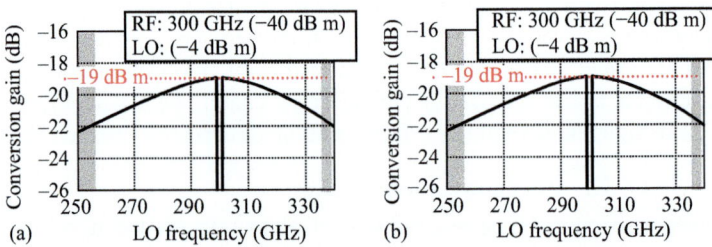

Figure 4.52   Simulated conversion gain (a) of the downconversion mixer with RF frequency at 300 GHz vs LO frequency with -4-dB m LO power, and output power of LO multiplier chain (b) when driven by 2-dB m $LO^{1/6}$ power. Copyright 2017 IEEE. Reproduced, with permission, from [45]

4.2.1.3   Measurement results

The RX was designed for the Taiwan Semiconductor Manufacturing Company (TSMC) Limited 40-nm 1P10M CMOS GP process. The chip area is $2.1 \times 1.5\,mm^2$ (Figure 4.53), and the chip is mounted on a PCB for measurement. DC power is supplied through bond wires. The power consumption is 0.65 W. High-frequency signals are supplied via RF probes. The $LO^{1/6}$ and RF signals are both generated by a VNA and the latter is upconverted by a WR3.4-band frequency extender. The downconverted signal is measured using a spectrum analyzer or a VNA. One of the differential IF outputs is terminated with a 50-Ω load because an external balun was not available.

Figure 4.54 shows measured and simulated output power and conversion gain as functions of RF input power. RF frequency is 294 GHz with $LO^{1/6}$ of 50.3 GHz. It corresponds to an IF output of LSB at 8 GHz. The simulated and measured output power are in agreement and show good linearity. The simulated 1-dB input compression point is 3.3 dB m. Figure 4.55 shows the RF frequency dependence of the conversion gain and NF. NF measurement is performed by using the cold-source

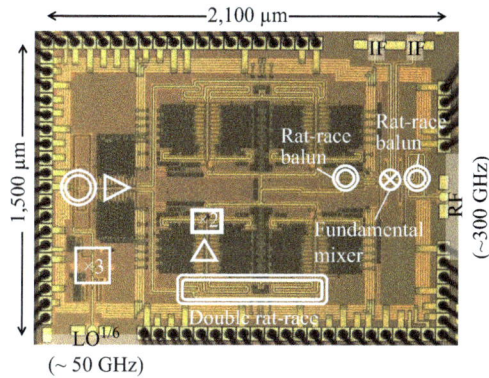

Figure 4.53   Die micrograph of the 300-GHz CMOS receiver. Copyright 2017 IEEE. Reproduced, with permission, from [45]

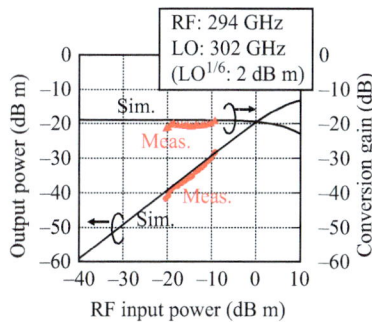

Figure 4.54   Output power and conversion gain with an IF output of LSB 8 GHz signal vs RF input power. $LO^{1/6}$ power of 2 dB m is supplied. Copyright 2017 IEEE. Reproduced, with permission, from [45]

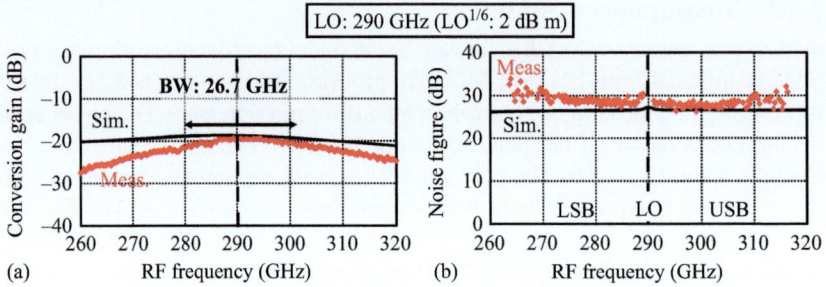

Figure 4.55   *(a) Conversion gain and (b) noise figure vs RF frequency. The $LO^{1/6}$ signal with a frequency of 48.3 GHz and a power of 2 dB m is applied. Copyright 2017 IEEE. Reproduced, with permission, from [45]*

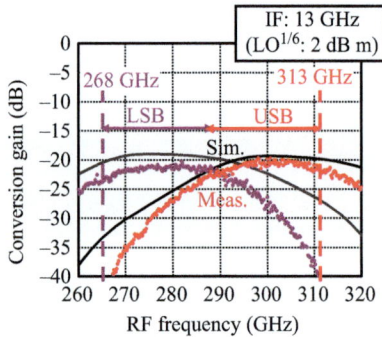

Figure 4.56   *Conversion gain with simultaneous change of RF and LO frequencies while keeping IF frequency fixed at 12.96 GHz. $LO^{1/6}$ power is 2 dB m. Copyright 2017 IEEE. Reproduced, with permission, from [45]*

method [41]. The peaks of the measured conversion gain and NF are -19.5 dB m and 27 dB, respectively. The 3-dB bandwidth of the measured conversion gain is 26.7 GHz and is narrower than the simulated result. It seems to have been caused by small mismatches in the LO multiplier chain. Figure 4.56 shows the conversion gain with simultaneous change of RF and LO frequencies while keeping IF frequency fixed at 12.96 GHz. If the LSB is used at lower frequencies and the USB is used at higher frequencies, the frequency range from 268 to 313 GHz can be covered by the RX. This indicates that the RX has a potential to utilize the vast frequency band above 275 GHz [30,36].

Figure 4.57 shows the measurement setup for short-range wireless communication. A TX test chip which has a capacity of upconverting a 105-Gbit/s QAM signal to the RF frequency range around 300 GHz is used [40]. A pair of horn antennas separated by 1 cm are used for the free space propagation of RF signal. The received RF signal at a center frequency of 294 GHz is fed to the RX chip via an RF probe. The $LO^{1/6}$ signal has a frequency of 50.3 GHz and a power of 2 dB m. Downconverted LSB signals at a center frequency of 8 GHz are measured using a vector signal

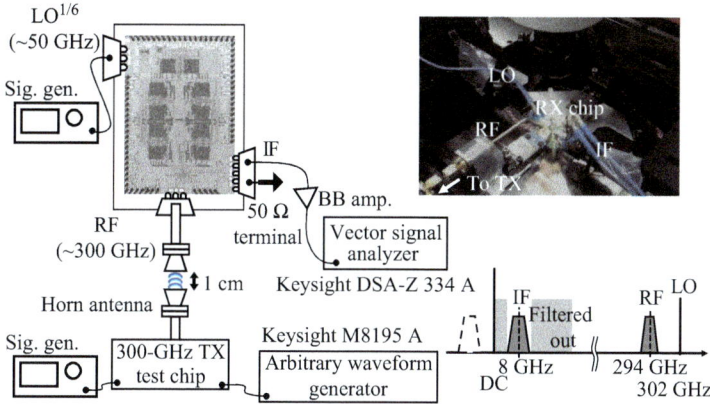

Figure 4.57   Measurement setup for wireless demonstration of RX. Copyright 2017
            IEEE. Reproduced, with permission, from [45]

Table 4.3   Signal constellations, error-vector magnitudes, symbol rates, and data
            rates

	QPSK	16QAM	32QAM
Constellation			
EVM	19.0%rms	12.2%rms	8.8%rms
BER	7.1×10^{-8}	9.3×10^{-5}	1.3×10^{-4}
Sym. rate	14 Gbaud	8 Gbaud	4 Gbaud
Data rate	28 Gb/s	32 Gb/s	20 Gb/s

analyzer. Table 4.3 shows the measured signal constellations, EVMs, symbol rates, and data rates. Data rates of 28, 32, and 20 Gbit/s are achieved with QPSK, 16QAM, and 32QAM with EVMs of 19.0%rms, 12.2%rms, and 8.8%rms, respectively.

Table 4.4 compares state-of-the-art THz RXs. Our RX might be the first CMOS RX to cover frequencies above 275 GHz.

4.2.2   Tripler-last LO driver [42]

In terahertz RXs, the power consumption of an LO multiplier consumes a large amount of power to obtain high conversion gain. An LNA-less 300-GHz CMOS RX using a doubler-last LO multiplier was discussed in Section 4.2.1 [45]. However, its power consumption was as high as 650 mW. In this section, another LNA-less 40-nm CMOS 300-GHz RX operating above f_{max} is discussed. It adopts a tripler-last LO multiplier and consumes 416 mW.

Table 4.4   Performance summary and comparison

	[27]	[7]	[39]	[43]	[44]	This work [45]
Technology	35 nm GaAs	130 nm SiGe	130 nm SiGe	90 nm CMOS	65 nm CMOS	40 nm CMOS
Freq. (GHz)	240	240	320	200	240	290
3-dB BW (GHz)	24	18	13	3	–	26.5
Conv. gain (dB)	3	11	−14	6.6	25	−19
NF (dB)	10	16	36	29.9	15	27
Modulation	8PSK	16QAM/ 64QAM	–	BPSK/ QPSK	BPSK/ QPSK	16QAM/ 32QAM
Data rate (Gbit/s)	96	0.7/1.0	–	4/2	10/16	32/20
P_{DC} (W)	–	0.87	0.216	0.063	0.26	0.65
Chip size (mm^2)	–	1.57	0.92	0.375	2	3.15
Integration	–	Ant., LNA, IQ mixer LO chain	SHM, LO chain	Mixer, IF amp.	Ant., Mixer IF amp., LO chain	Mixer, LO chain

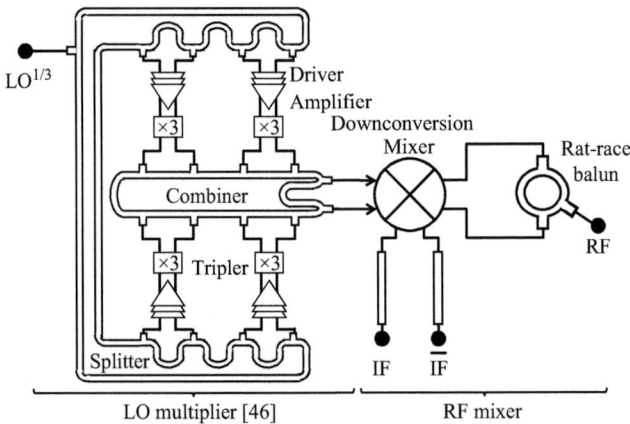

Figure 4.58   Schematic diagram of a 300-GHz receiver. Copyright 2017 IEEE. Reproduced, with permission, from [42]

Figure 4.58 shows the schematic diagram of our 300-GHz RX. The circuit consists of a 300-GHz downconversion mixer and an LO multiplier. A W-band LO signal at around 100 GHz (LO$^{1/3}$) is supplied from an off-chip signal source. The downconversion mixer outputs an IF signal at several GHz. The mixer part is composed of a balun and a double-balanced fundamental passive mixer consisting of four 32-μm-wide transistors shown in Figure 4.59. The LO and RF signals are applied to their gates and drains, respectively. The matching networks are implemented using open and shorted stubs. The LO multiplier consists of a rat-race-based 8-way power splitter, W-band driver amplifiers, triplers upconverting fundamental frequency into

Double-balanced
fundamental mixer

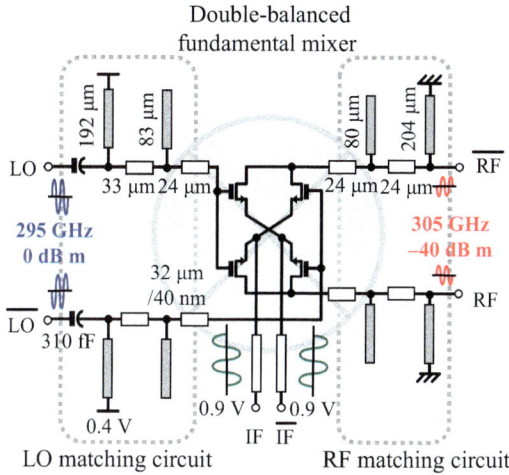

*Figure 4.59   Schematic diagram of a downconversion mixer used in the 300-GHz
receiver. Copyright 2017 IEEE. Reproduced, with permission,
from [42]*

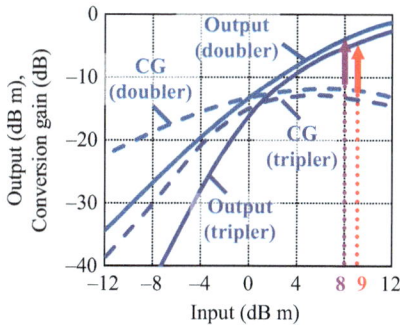

*Figure 4.60   Output power and conversion gain of a differential tripler and
single-ended doubler upconverting into 300 GHz vs input power. The
input frequencies of the tripler and the doubler are 100 and 150 GHz,
respectively. Copyright 2017 IEEE. Reproduced, with permission,
from [42]*

300 GHz, and a rat-race-based power combiner. The output power of the LO multiplier is about 2 dB m at 302 GHz. Figure 4.60 shows simulated performance of a
300-GHz differential doubler [45]. The output power of the doubler is higher than
that of the tripler by only 1.2 dB m above the compression point. Generally speaking,
a doubler has higher performance compared with a tripler. However, in this case,
the input frequencies are different, that is, 100 GHz for the tripler and 150 GHz for
the doubler. Therefore, the performance advantage of the doubler is not very significant. To supply an 8 dB m power to a 100-GHz-input tripler and a 150-GHz-input
doubler, 3- and 6-stage cascode driver amplifiers are needed, respectively. The power

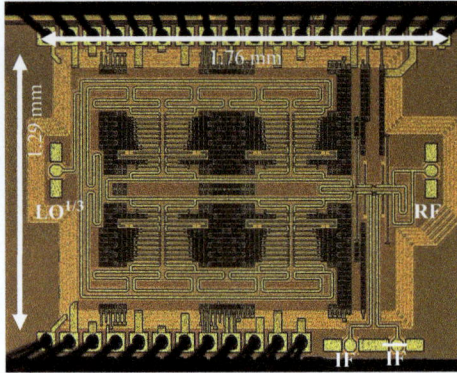

Figure 4.61   *Die micrograph of the 300-GHz CMOS receiver. Copyright 2017 IEEE. Reproduced, with permission, from [42]*

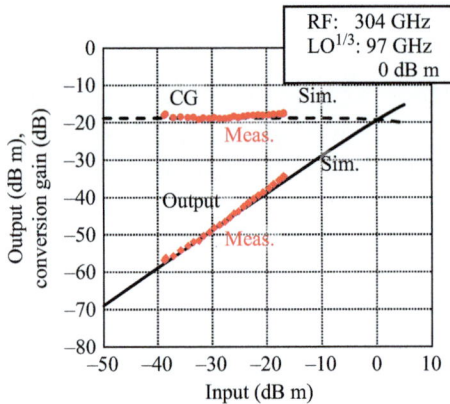

Figure 4.62   *Output power and conversion gain with an IF output of LSB 13 GHz signal vs RF input power. $LO^{1/3}$ power of 0 dB m is supplied. Solid lines and dots show simulated and measured results. Copyright 2017 IEEE. Reproduced, with permission, from [42]*

consumption of these is about 81 and 168 mW, respectively. Based on the above, the tripler-last LO multiplier has an advantage in terms of power consumption because more efficient driver amplifier can be designed at 100 GHz.

4.2.2.1   Measurement results

The RX was designed for the TSMC 40-nm 1P10M CMOS GP process. The chip area is $1.76 \times 1.29\ mm^2$ (Figure 4.61) and the chip is mounted on a PCB for measurement. DC power is supplied through bond wires. The power consumption is 416 mW. High-frequency signals are supplied via RF probes. The $LO^{1/3}$ and RF signals are generated by a signal generator with a WR10-band source module and a VNA with a WR3.4-band frequency extender, respectively. The downconverted signal is measured using

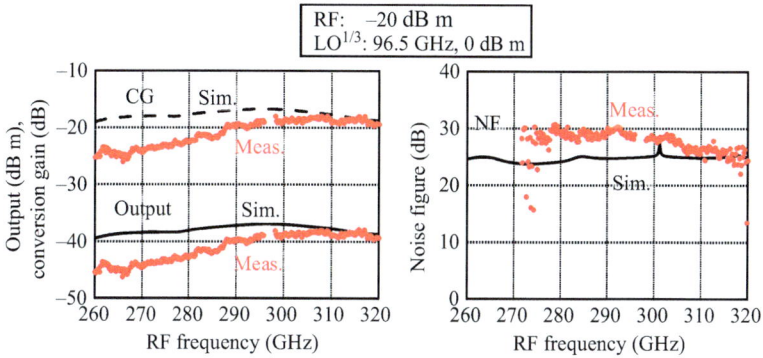

Figure 4.63 Output power, conversion gain and noise figure vs RF frequency. The LO$^{1/3}$ signal power is 0 dB m at a frequency of 96.5 GHz. Solid lines and dots show simulated and measured results. Copyright 2017 IEEE. Reproduced, with permission, from [42]

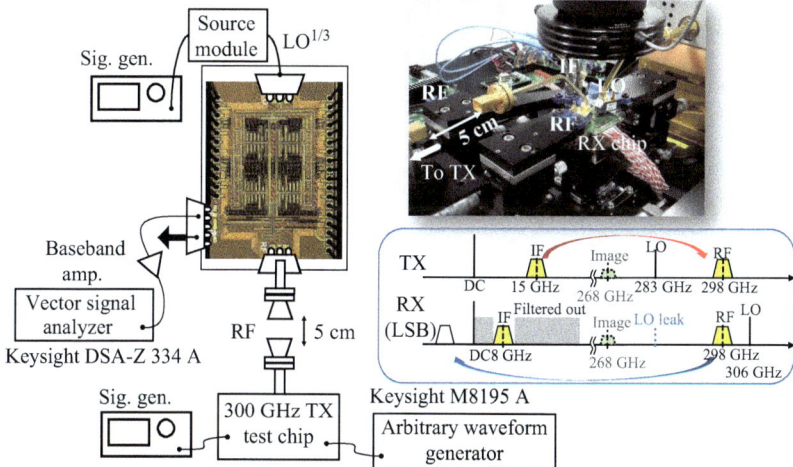

Figure 4.64 Measurement setup for wireless demonstration of receiver. Copyright 2017 IEEE. Reproduced, with permission, from [42]

a spectrum analyzer or a VNA. One of the differential IF outputs is terminated with a 50-Ω load. Figure 4.62 shows measured and simulated output power and conversion gain as functions of RF input power. The RF frequency is 304 GHz and LO$^{1/3}$ is 97 GHz. It gives an IF output of the USB at 13 GHz. The simulated and measured output powers are in agreement and show good linearity. The conversion gain and NF are shown in Figure 4.63 and are −18 dB m and 25.5 dB, respectively. The 3-dB bandwidth is about 33 GHz at a center frequency of 308 GHz. Figure 4.64 shows the measurement setup for short-range wireless communication. A TX test chip capable of upconverting a QAM signal to the RF frequency range around 300 GHz is used, which

Table 4.5   Signal constellations, error-vector magnitudes, symbol rates, and data rates

	QPSK	16QAM	32QAM
Constellation			
EVM	18.6%rms	13.4%rms	8.54%rms
BER	3.8×10^{-8}	3.2×10^{-4}	8.3×10^{-5}
Sym. rate	10 Gbaud	8 Gbaud	4 Gbaud
Data rate	20 Gb/s	32 Gb/s	20 Gb/s

Table 4.6   Performance summary and comparison

	[27]	[7]	[39]	[44]	[45]	This work
Technology	35 nm GaAs	0.13 μm SiGe	0.13 μm SiGe	65 nm CMOS	40 nm CMOS	40 nm CMOS
Freq. (GHz)	228–252	227–245	315–328	240	280–307	287–320
Conv. gain (dB)	3	11	−14	25	−19	−18
NF (dB)	10	16	36	15	27	25.5
Modulation	8PSK	QPSK/ 16QAM/ 64QAM	–	BPSK/ QPSK	QPSK/ 16QAM/ 32QAM	16QAM/ 32QAM
Data rate (Gbit/s)	96	2.7/0.7/1.0	–	10/16	28/32/20	32/30
P_{DC} (W)	–	0.87	3.072	0.26	0.65	0.416
Chip size (mm^2)	–	1.57	0.92	2	3.15	2.29
Integration	–	Antenna LNA IQ mixer LO chain	SHM LO chain	Antenna Mixer IF amp. LO chain	Mixer LO chain	Mixer LO chain

can cancel LO leak and suppress the images [40]. A pair of horn antennas separated by 5 cm are used for the free space propagation of the RF signal. The received RF signal at a center frequency of 298 GHz is fed to the RX chip via an RF probe. The LO$^{1/3}$ signal is supplied to the RX chip at a frequency of 102 GHz with a power of 3 dB m. The downconverted signal at a center frequency of 8 GHz is measured using a vector signal analyzer. Table 4.5 shows the measured signal constellations, EVM, symbol rates, and data rates of LSB signals. Data rates of 32 and 20 Gbit/s are achieved with 16QAM and 32QAM with EVMs of 13.4%rms and 8.6%rms, respectively. Table 4.6

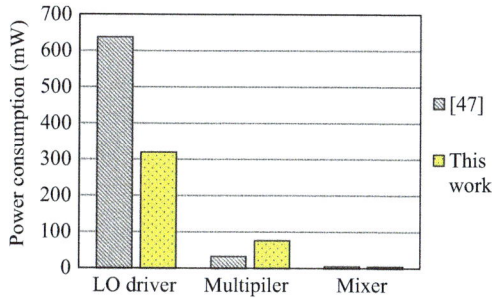

Figure 4.65   Power breakdowns of terahertz receivers. Copyright 2017 IEEE. Reproduced, with permission, from [42]

compares state-of-the-art terahertz RXs. High performance comparable to [45] is achieved with much lower power consumption (shown in Figure 4.65) owing to the high performance tripler-last LO multiplier.

4.2.3   Summary

Two types of RXs with CMOS 40 nm process are discussed. The difference is that LO driver adopts either doubler or tripler architecture. At the TX, the maximum output power of the doubler architecture is larger than that of the tripler architecture. However, to obtain the same output frequency, the input frequency of the tripler can be lower than that of the doubler. For this reason, in an RX using the triple-last LO driver, power consumption is reduced while obtaining equivalent performance. Both RXs have a maximum data rate of 32 Gbit/s, which is lower than that of the TX. The primary reason for this is that raising the output power of the TX directly relates to the parallel number of last-stage mixers, but lowering the NF of the RX is not directly related to parallel number of the LO drivers. No matter how much the output power of the LO driver is raised, the improvement of NF will hit a peak. For this reason, NF cannot be unlimitedly improved by merely increasing the output power of the LO driver. This is a crucial factor that there is no significant difference in characteristics even if the LO driver is changed from doubler- to tripler architecture. (The performance of the circuit using the tripler-last LO driver was a little preferable!) To improve NF with the LNA-less architecture, it is necessary to further reduce the conversion loss of the mixer, and to search for the condition that the NF becomes small in the terahertz band will be necessary. Research on RXs has more issues than that on TXs.

4.2.4   Receiver module [48]

The currently unallocated terahertz (THz) frequencies above 275 GHz have been considered as a promising platform for ultrahigh-speed broadband wireless communications [36,49]. The physical-layer challenges that lie ahead include not only the design of THz TX and RX integrated circuits but also their packaging. Recently, a 300-GHz Si CMOS RX was reported, which operated above the unity-power-gain frequency, f_{max}, of the NMOSFET [45]. The RX, together with a CMOS TX [31], achieved a wireless data rate of 32 Gbit/s with 16QAM in 300-GHz band. However, the experiment was performed by using a WR3.4 WG probe, without packaging the RX chip. For practical use, the CMOS RX chip must be packaged and modularized.

To compensate for relatively high atmospheric propagation losses at such high frequencies, the size of antennas used for wireless communications must be large. A rectangular WG interface offers the possibility of using a choice of high-gain antennas and off-chip amplifiers. However, realizing direct transition from a chip to a WG is difficult; it requires precise cutting and back-grinding of a chip. It seems an ideal and simple method for THz CMOS technology that a mounted chip is connected with a WG transition on a board.

In recent years, active development of packaging techniques for THz electronics have been undertaken [29,33,50]. A microstrip-to-WG transition has been reportedly integrated in a multilayer LTCC substrate for 300 GHz [33]. The design utilized a vertical hollow WG with a back-short structure located inside a multilayer LTCC board. More recently, the details of transmission-line-to-WG transition with a back-short structure built directly into a glass epoxy PCB were reported [51], wherein the CMOS TX chip was mounted onto the PCB using flip-chip bonding. Glass epoxy is more suitable for mass production than LTCC due to its low cost, despite having a higher loss tangent than LTCC. Although the TX module that utilized the multilayered glass epoxy PCB showed somewhat higher packaging losses, it achieved a high wireless data rate of 48 Gbit/s.

This study presents the performance characteristics of a CMOS RX module with a CMOS-chip-to-WG transition integrated into a low-cost multilayered glass epoxy PCB.

4.2.4.1   300-GHz CMOS receiver module

Figure 4.66(a) illustrates the layer structure of the glass epoxy PCB for the designed RX module. The configuration is similar to that of the previously described CMOS TX module [51]. Gold stud bumps with a flip-chip bonding mechanism were used to mount the CMOS RX chip underneath a WR3.4 WG flange near the center of the PCB. The LO signal, at a frequency of approximately 50 GHz, can be supplied from an external source via a coaxial connector that is mounted on the south edge of the PCB [not shown in Figure 4.66(a)]. This LO signal is first upconverted into approximately 300 GHz via a 6× multiplier chain and subsequently fed to the downconversion mixer in the CMOS RX chip. The IF signal at the frequency of several GHz, which is downconverted by the mixer, leaves the PCB via coaxial connectors mounted on the north edge of the PCB. GCPWs are used for the signal paths between the coaxial connectors and stud bumps. A magnified view of the GCPW-to-rectangular-WG transition is shown in Figure 4.66(b). Herein, the pads for the 300-GHz RF signal are 75-μm-pitch GSG pads, and the height of the bumps is about 20 μm. The GCPW-to-WG transition has a vertical hollow cut into the PCB's four metal layers, serving as a back short. The ideal design depth for the back-short structure is $\lambda/4$, where λ represents the RF signal wavelength. Considering the nominal relative permittivity of the glass epoxy board at 10 GHz and together with the restriction on the feasible depths by the multilayered structure, the depth of 135 μm was selected for the back-short

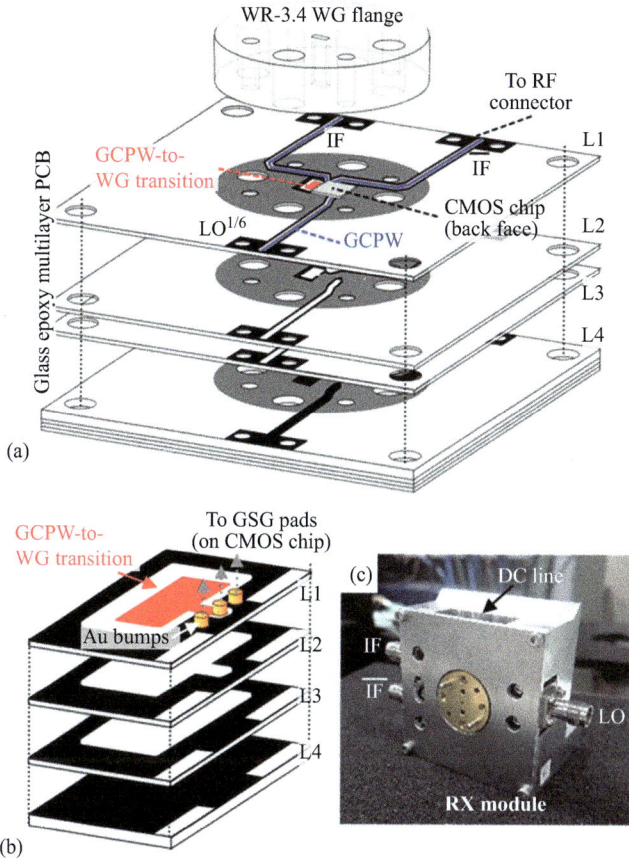

Figure 4.66 Structure of (a) the RX module PCB and (b) a close-up view of the WG transition at the center. (c) Photo of the receiver module. Copyright 2018 IEEE. Reproduced, with permission, from [48]

structure in the studied module. An EM field simulation was conducted using ANSYS HFSS, which resulted in a nominal glass epoxy loss tangent of 0.0076 at 10 GHz, indicating that the insertion loss of the near-$\lambda/4$ WG (back short) is approximately 0.08 dB at 300 GHz. The dimensions of the rectangular probe, comprising the PCB's top metal layer in Figure 4.66(b) (shown in red), were selected to achieve an impedance matching the impedance induced between the transmission line (GCPW) and the WR3.4 rectangular WG. The dimensions of the vertical hollow WG cross section were similarly adjusted for impedance matching, resulting in dimensions that were 0.8 times the inner dimensions of the WR3.4 WG. The results of the EM field simulation of the GCPW-to-WG transition also showed that the value of S_{21} is -0.23 dB and that S_{11} and S_{22} stay below -10 dB over a 45-GHz bandwidth. Figure 4.66(c) shows a photo of the RX module after the completion of assembly.

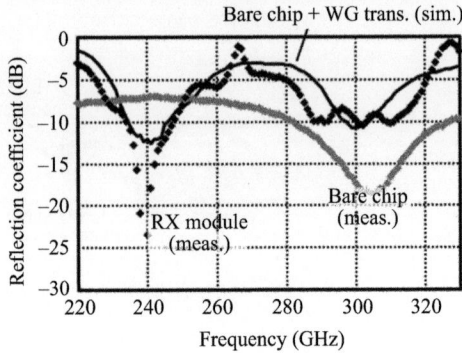

Figure 4.67   Reflection coefficients of the RF port of the module and the bare chip. Solid line corresponds to the simulated result using a more precise simulation model identified after fabrication. Copyright 2018 IEEE. Reproduced, with permission, from [48]

4.2.4.2   Measurement results

A VNA was used to measure the signal downconverted by the 300-GHz CMOS RX module. The RF signal, generated by the VNA with a WR3.4-band frequency extender, was thereupon supplied to the RX module via a one-inch-long straight WR3.4 WG with an insertion loss of approximately 1 dB. The $LO^{1/6}$ signal at a frequency around 50 GHz was supplied from a signal generator. One of the differential IF output ports was terminated with a 50-Ω load. Measured and simulated reflection coefficients at the RF port of the RX module are shown in Figure 4.67. A subsequent post-fabrication simulation was performed using a more precise model than a prefabrication counterpart and shows a fairly good agreement with the measured reflection coefficient of the RX module. Given the fact that the reflection coefficient of the module is significantly higher than that of the bare chip, the performance of the RX module could be improved if the result of the post-fabrication simulation is reflected to future design of a WG transition.

The measured output power and conversion gain as a function of the input RF power is shown in Figure 4.68. The RF frequency was 291 GHz, and the $LO^{1/6}$ was at 48.3 GHz and its power was 2 dB m. The peak conversion gain was about −23.7 dB m. This result indicates that a packaging loss of about 4 dB, as the bare CMOS chip, had an output RF power of −19.5 dB m.

Figure 4.69 shows the frequency response of the conversion gain and NF of the module and bare chip. The RF power was −15 dB m, and the $LO^{1/6}$ signal was the same as in Figure 4.68. The 3-dB bandwidth of the module was about 18.4 GHz. Compared with the result of the bare chip, the conversion gain of the module decreased sharply at the frequency above 300 GHz. The NF of the module was about 33 dB, which is higher than that of bare chip by 6 dB. The deterioration of the NF was slightly larger than that of the conversion gain by 2 dB. A possible reason is a measurement error due to calibration fault of the NF measurement system. Another possibility is that

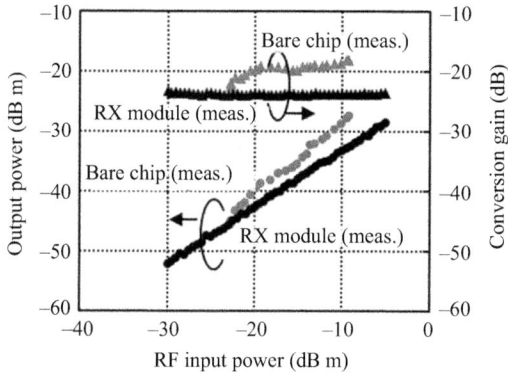

*Figure 4.68   Measured output power and conversion gain of the RX module and
bare chip with an IF output as a function of RF input power. The RF
frequency was 291 GHz. The LO$^{1/6}$ signal was applied at 48.3 GHz
and 2 dB m power. Copyright 2018 IEEE. Reproduced, with
permission, from [48]*

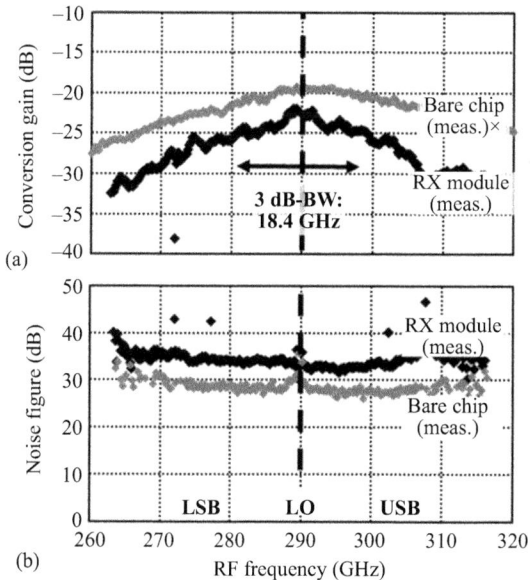

*Figure 4.69   Measured conversion gain (a) and noise figure (b) of the RX module
and of the bare chip as functions of the RF input power. The RF
frequency was 291 GHz. The LO$^{1/6}$ signal was applied at 48.3 GHz
and 2 dB m power. Copyright 2018 IEEE. Reproduced, with
permission, from [48]*

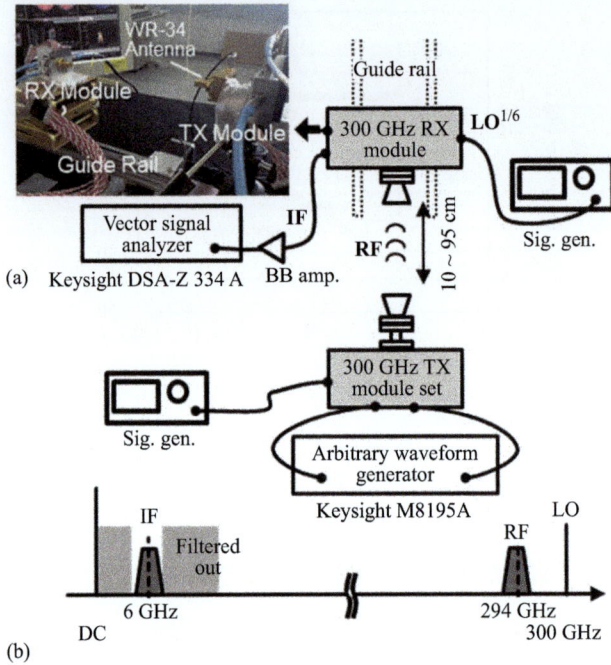

Figure 4.70 (a) Conversion gain and (b) noise figure of the RX module and bare chip vs RF frequency. The RF power was −15 dB m. The LO$^{1/6}$ power was 2 dB m at 48.3 GHz. Copyright 2018 IEEE. Reproduced, with permission, from [48]

the source impedance shifted in such a way that the NF was affected more than the conversion gain.

Figure 4.70 shows the measurement setup for evaluating the wireless performance of the RX module. A TX module capable of upconverting a 48-Gbit/s QAM signal to the RF frequency around 300 GHz was used [51]. A pair of WR3.4 standard horn antennas with a gain of 24 dB i were used. The received RF signal at a center frequency of 294 GHz was fed to the RX module via the horn antenna. A downconverted LSB signal at a center frequency of 6 GHz was measured using a vector signal analyzer via an external baseband (BB) amplifier with a gain of about 30 dB. The RX module was put on a sliding mount on a guide rail for examining the distance dependence.

Figure 4.71 shows the EVM of QPSK and 16QAM as a function of the distance. The EVMs increased proportionally at the distances of above 20 cm, which appeared to be a boundary of the antenna's near and far field. The highest data rate and maximum transmission distance with (bit-error rate) BER of below 10^{-4} were 20 Gbit/s with 16QAM at the distance of 10 and 75 cm at a data rate of 2 Gbit/s with QPSK, respectively (shown in Table 4.7).

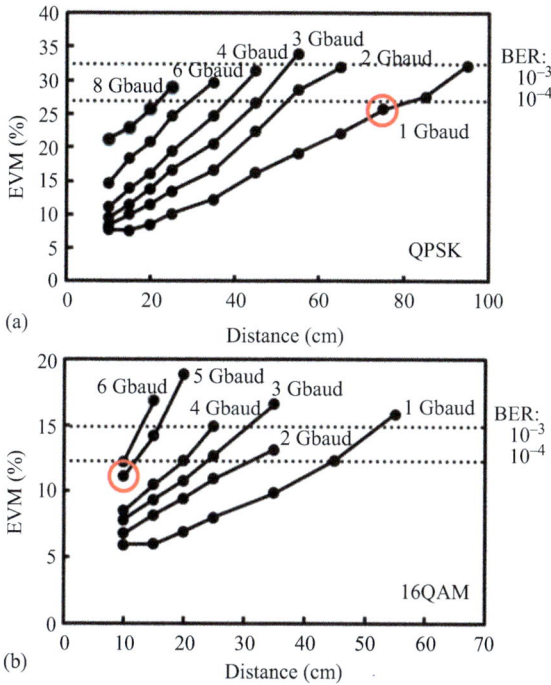

Figure 4.71   Measured error-vector magnitude of (a) QPSK and (b) 16QAM as a function of distance. Copyright 2018 IEEE. Reproduced, with permission, from [48]

Table 4.7   Signal constellations, distance, error-vector magnitudes, symbol rates, and data rates with a maximum transition distance and highest data rate with BER of below 10^{-4}

	QPSK	**16QAM**
Constellation		
Distance	75 cm	10 cm
EVM	25.7%rms	12.2%rms
BER	5.1×10^{-5}	2.2×10^{-5}
Sym. rate	1 Gbaud	5 Gbaud
Data rate	2 Gb/s	20 Gb/s

4.2.4.3   Conclusion

This study presents the design and analysis of a 300-GHz CMOS RX module with a CMOS-chip-to-WG transition built into a multilayered glass epoxy PCB. The CMOS RX chip was mounted on the PCB using a gold stud bump flip-chip bonding method. A back-short structure was used for the transmission-line-to-WG transition and constructed using a vertical hollow WG structure inside the four-layer PCB. The measured conversion gain, NF, and 3-dB bandwidth of the RX module are −23.7 dB, 33 dB, and 18.4 GHz, respectively. A wireless data rate of 20 Gbit/s was achieved using the modules with 16QAM. Possible performance improvements for the RX module could be achieved using post-fabrication information obtained from the simulation results. Additionally, the use of WG flange with an improved design [52] could give better performance and measurement repeatability.

4.3   One-chip transceiver [53]

IEEE Standard 802.15.3d, published in October 2017, defines a high-data-rate wireless physical layer that enables up to 100 Gbit/s using the lower THz frequency range between 252 and 325 GHz (hereafter referred to as the "300-GHz band"). It stipulates that the 300-GHz band be channelized into thirty-two 2.16-GHz-wide channels (Figure 4.72) or a smaller number of wider channels whose bandwidths are all integer multiples of 2.16 GHz. This section presents a CMOS transceiver (TRX) chip targeted at channels 49 through 51 and 66 of 802.15.3d (Figure 4.72). There have been reports on solid-state TRXs operating in or near the 300-GHz band [6,11,29,54–56]. Some of these [29,54,55] were TX/RX or block-level chipsets, which can enjoy more flexibility in design and independent optimization of TX and RX. They successfully achieved ≥ 64 Gbit/s. On the other hand, single-chip TRXs [6,11,56] did not always reveal achievable data-rates and nor were they capable of supporting QAM. Nevertheless, eventual development of full-featured single-chip TRXs is desirable especially for applications requiring deployment of many TRXs, as is envisioned implicitly by 802.15.3d. The single-chip QAM-capable CMOS TRX presented herein is an outcome of the efforts in that direction.

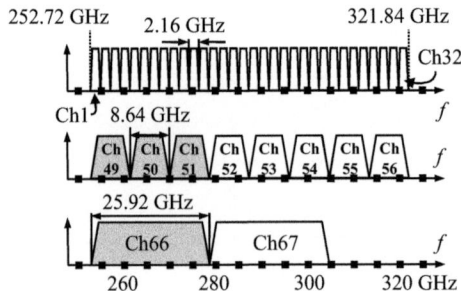

Figure 4.72   300-GHz-band channel allocation published by IEEE 802.15.3d. Channels 49–51 and 66 are used in the transceiver. Copyright 2019 IEEE. Reproduced, with permission, from [53]

4.3.1 Architecture

Since the 40 nm CMOS process that we used has a relatively low f_{max}, we adopt PA- and LNA-less architecture. A simplified schematic of the TRX is shown in Figure 4.73. It operates either in TX or RX mode shown in Figure 4.74. These modes share the TX part, which serves as an LO multiplier chain in RX mode. The TX mode is architecturally similar to the mixer-last TX in [40], but the input is BB as opposed to IF. The input signal is upconverted by a quadrature modulator and the IF is superposed on LO ($\sim$ 133 GHz) by the modulator itself. The resulting (LO $\pm$ IF) are squared by frequency doublers (or square mixers [40]), producing (LO $\pm$ IF)^2. (LO $+$ IF)^2 and (LO $-$ IF)^2 are then fed, respectively, to the positive ($+$) and negative ($-$) ports of a rat-race balun variant (double-rat-race [40]), and the difference (Δ) port outputs the desired RF$_{TX}$ signal, 4LO $\cdot$ IF($\sim$ 266 GHz). In RX mode, the BB input ports

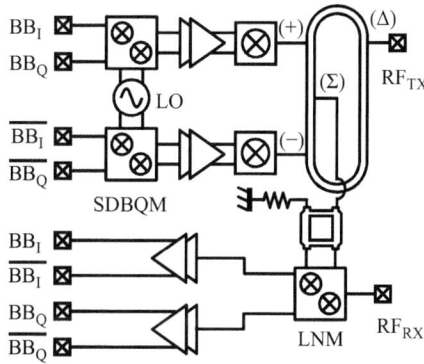

Figure 4.73   300-GHz-band CMOS one-chip transceiver architecture. Copyright 2019 IEEE. Reproduced, with permission, from [53]

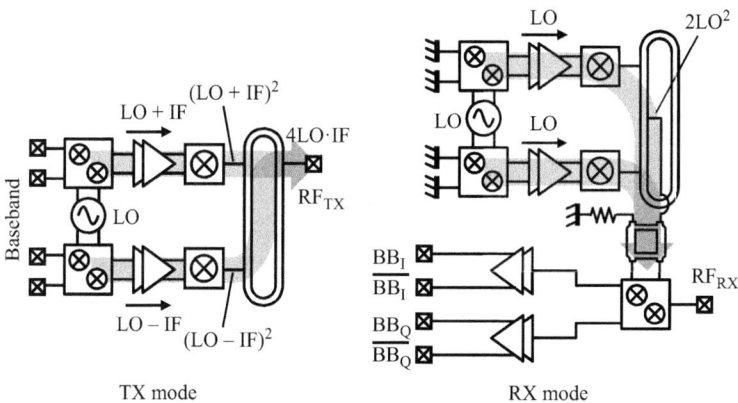

Figure 4.74   Signal flow in transmitter (TX) and receiver (RX) modes of 300-GHz-band CMOS transceiver. Copyright 2019 IEEE. Reproduced, with permission, from [53]

are terminated appropriately and the modulator works only as an LO buffer. The doublers generate LO^2 and the sum (Σ) port of the double-rat-race feeds the RX part with $2LO^2 (\sim 266\,\text{GHz})$. The mixer-first direct-conversion RX part consists of a low-noise mixer (LNM) and BB amplifiers.

4.3.2   Transmitter mode

The quadrature modulator in the shared TX part is required to leak a certain amount of LO to the output, and therefore well-balanced double-balanced mixers cannot be used. There exists an optimum leaked LO power that gives the highest RF power in TX mode. To make the amount of LO leakage controllable, we use what could be called a "semidoubly balanced quadrature mixer," shown in Figure 4.75. It is schematically similar to a double-balanced mixer but BB inputs are not balanced. What would have been the "BB inverse ports" are used to tune the coefficient α in the output, $(\alpha LO + IF)$. Different values of α are chosen in TX and RX modes. The last stage of the IF amplifier is a stacked doubler driver having large output voltage amplitude (Figure 4.76). This configuration contributes significantly to high RF (LO) power in TX (RX) mode. The stacked driver and the doubler were optimized together

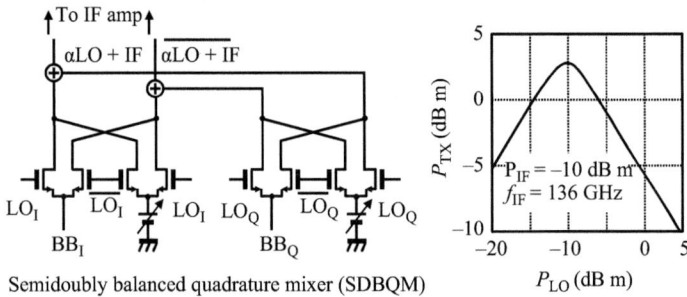

Figure 4.75   *Semidoubly balanced quadrature mixer (SDBQM) used in the first up-conversion mixer in TX mode. Copyright 2019 IEEE. Reproduced, with permission, from [53]*

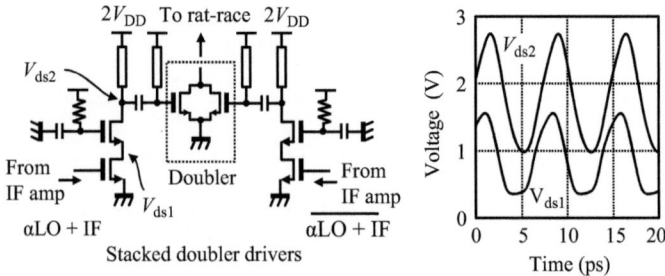

Figure 4.76   *Stacked doubler driver. To improve the output power of the doubler, it is necessary to increase the input power of the doubler. Copyright 2019 IEEE. Reproduced, with permission, from [53]*

as a single block. This split of the IF amplifier into the doubler driver and the rest facilitated wideband design.

4.3.3 Receiver mode

Noise performance is of utmost importance for the dedicated RX part. However, conversion gain of the downconversion mixer cannot be greater than unity (0 dB) because of low $f_{\max}$. Clearly, the noise factor and conversion loss of the mixer must both be as low as possible, but performance metrics that require bona fide gain (>1), such as the noise measure, do not help quantify and compare performance of various mixer configurations. We arrived at the somewhat ordinary LNM configuration, shown in Figure 4.77, by looking at the Engberg–Gawler factor given by

$$\text{EG} = \left(1 - \frac{1}{G_1}\right)\left(1 - \frac{M_1}{F_2 - 1}\right), \tag{4.9}$$

where G_1 and M_1 are the available (conversion) gain and the (SSB) noise measure of the first stage of two cascaded stages, respectively, and F_2 is the noise factor of

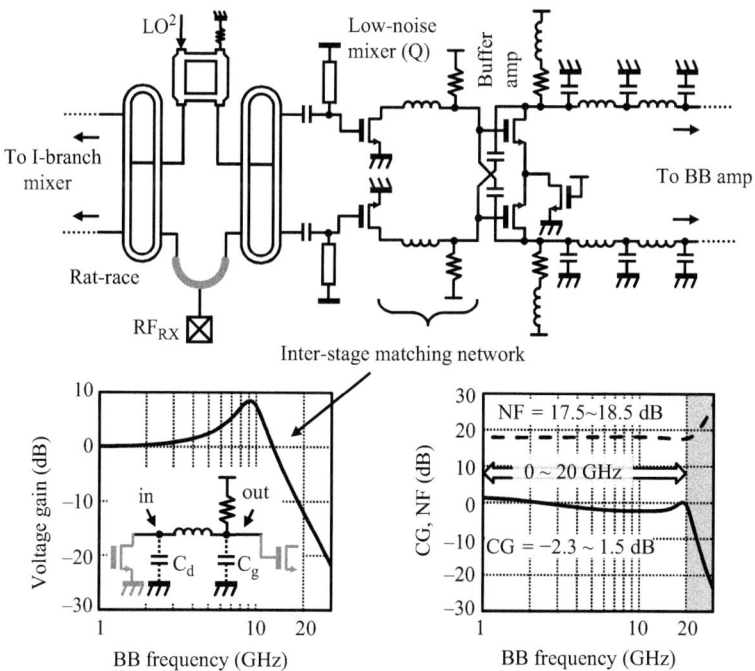

Figure 4.77 Schematic of low-noise mixer and the first stage of baseband (BB) amplifier. By adjusting the parameters of the interstage matching network, the reduction of the NF and the flat frequency characteristic are realized. Copyright 2019 IEEE. Reproduced, with permission, from [53]

the second stage [57]. In our case, the "first stage" includes the mixer and a buffer amplifier (Figure 4.77), and the "second stage" is the BB amplifier. The latter was optimized independently of the "first stage" and therefore F_2 has a fixed value. EG serves as a useful performance metric even when $G_1 < 1$ and $M_1 < 0$: the larger (more to the right on a real number line) the value of EG, the better. The LO signal, LO^2, for downconversion coming from the TX part and RF_{RX} from the receiving antenna port are combined by a rat race and applied to the MOSFET gates of the "first stage." Simulated NF and conversion gain of the "first stage" are fairly flat over a 20 GHz BB bandwidth (Figure 4.77). This was accomplished by high-frequency gain peaking of the interstage matching network (Figure 4.77).

4.3.4   Schematic

Figure 4.78 shows the schematic of the entire TRX. LO_{IN} is supplied from an external signal generator. After being split into in-phase and antiphase branches, it goes through frequency triplers before reaching the quadrature modulators (SDBQMs). Branch-line quadrature hybrids are used to produce quadrature phase in TX and RX parts. The four square mixers perform final upconversion. The double-rat-race power-combines RF and cancels out LO^2 in TX mode, whereas it power combines LO^2 in RX mode. Transmitting and receiving antennas are to be connected to RF_{TX} and RF_{RX} ports, respectively. The RX part has I and Q branches.

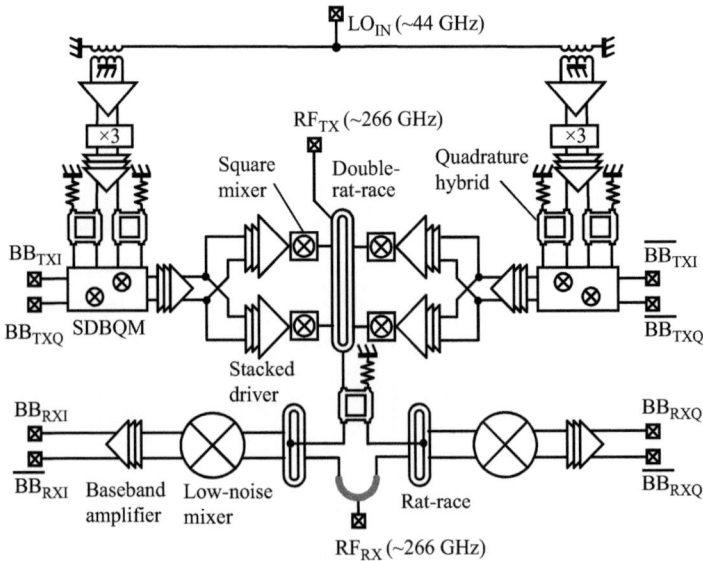

Figure 4.78   Overall schematic of the 300-GHz-band one-chip CMOS transceiver. Copyright 2019 IEEE. Reproduced, with permission, from [53]

4.3.5 Measurement

The TRX was fabricated using a 40 nm CMOS process. A chip micrograph of our TRX is shown in Figure 4.79. The die size is 4.92×2.25 mm^2. Figure 4.80 shows the TRX characterization setup and Figure 4.81 shows measurement results. The TRX chip was mounted on a PCB. LO$_{IN}$ and DC power were supplied to the chip through bond wires. BB input signals were generated by an arbitrary waveform generator and were fed through a multi-tip probe. TX output signal RF$_{TX}$ was led to a WR3.4-band block downconverter via a WG probe, and a spectrum analyzer measured the power. The frequency dependence of TX output power was measured by sweeping the frequency of the sinusoidal BB input at a fixed LO^2 frequency, f_{LO^2}. The input-power dependence of the TX output power was measured at a fixed BB frequency of 5 GHz. The observed LO leakage reduction can be understood analogously to "blocking" in an RX, in which a powerful interferer (IF in Figure 4.74 in this case) accompanying a

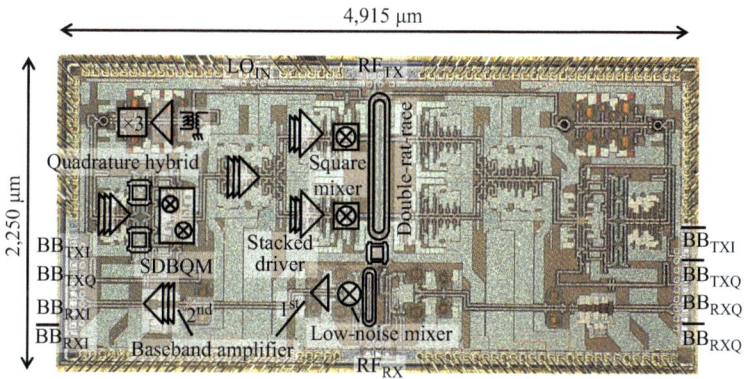

Figure 4.79   Chip micrograph. Copyright 2019 IEEE. Reproduced, with permission, from [53]

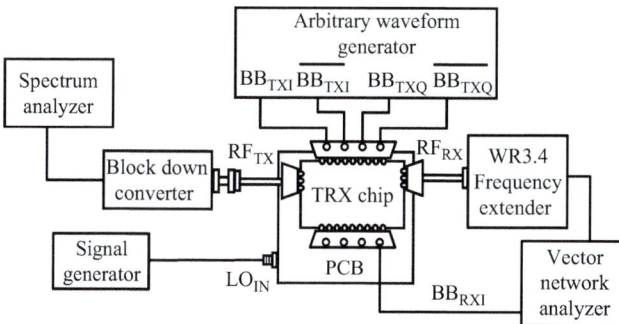

Figure 4.80   Measurement setup. Copyright 2019 IEEE. Reproduced, with permission, from [53]

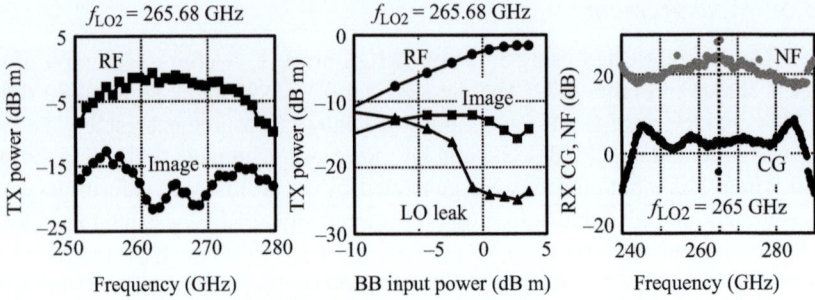

Figure 4.81   *Frequency responses and power responses in transmitter mode and frequency response in receiver mode. Copyright 2019 IEEE. Reproduced, with permission, from [53]*

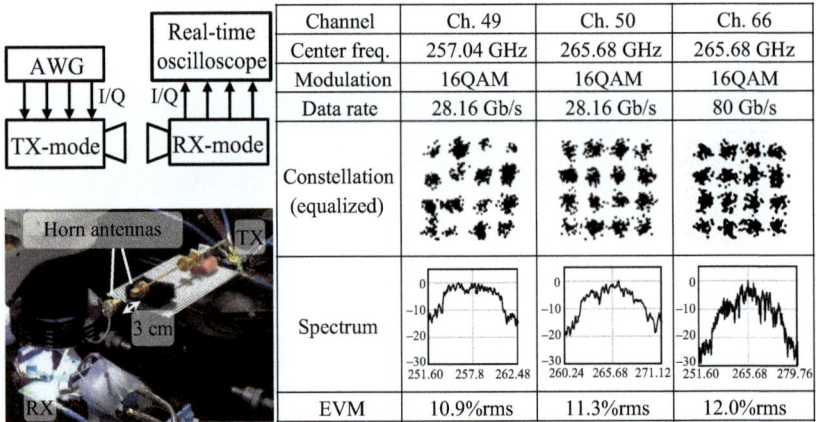

Channel	Ch. 49	Ch. 50	Ch. 66
Center freq.	257.04 GHz	265.68 GHz	265.68 GHz
Modulation	16QAM	16QAM	16QAM
Data rate	28.16 Gb/s	28.16 Gb/s	80 Gb/s
Constellation (equalized)			
Spectrum			
EVM	10.9%rms	11.3%rms	12.0%rms

Figure 4.82   *Measurement setup for communication experiment and measured constellations. Copyright 2019 IEEE. Reproduced, with permission, from [53]*

weaker desired signal (LO in Figure 4.74) causes gain compression. RX conversion gain and NF were measured using a multifunction VNA and a frequency extender through observing one of the four BB outputs. TX mode and RX mode consumed DC power of 890 and 897 mW, respectively.

Figure 4.82 shows the wireless link measurement setup. A pair of 24-dB i horn antennas were used. Output BB signals from the RX were analyzed using a real-time oscilloscope with vector signal demodulation and channel equalization capabilities. Signal constellations and power spectra are shown for frequencies and bandwidths corresponding to channels 49, 50, and 66 of 802.15.3d (Figure 4.82). A data-rate of 80Gbit/s over a distance of 3 cm was achieved with 16QAM. Recently reported THz TRXs are compared in Table 4.8.

Table 4.8   Performance comparison

	[29]	[54]	[55]	[56]	[6]	This work
Technology	35 nm	0.13 μm	80 nm	0.13 μm	250 nm	40 nm
	GaAs	SiGe HBT	InP	SiGe BiCMOS	InP	CMOS
Freq. (GHz)	300	240	287	340	298	265.68
Modulation	QPSK	QPSK	16QAM	FM/AM	–	16QAM
Single chip	N	N	N	Y	Y	Y
P_{out} (dB m)	−4	8.5	9.5	0.1	−2.3	−1.6
P_{DC} (W)	–	1.96	–	1.7	0.45	1.79
Data rate (Gbit/s)	64	65	100	–	–	80

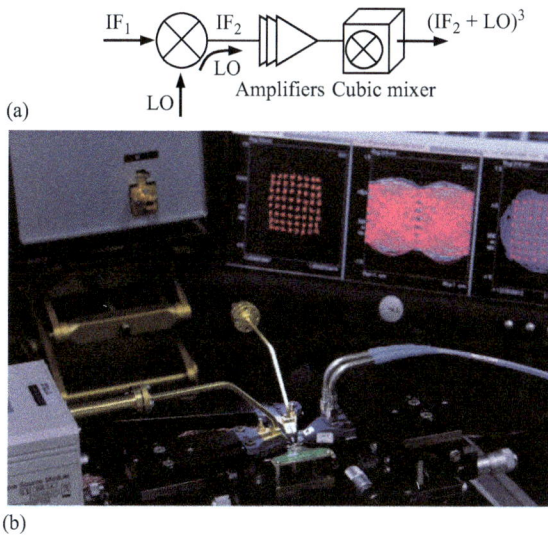

Figure 4.83   300-GHz CMOS transmitter: (a) simplified block diagram and (b) snapshot of wireless digital data transmission experiment. Copyright 2016 IET. Reproduced, with permission, from [25]

4.4   Wireless-link evaluation [25]

4.4.1   Introduction

The frequency band above 275 GHz currently remains unallocated, and its spectrum allocation is due to be discussed. Some wireless TRXs operating above 275 GHz have been realized with compound semiconductors [2,3] or photonic devices [35,46,58,59]. We recently reported a 300-GHz CMOS TX that operates above the transistor unity-power-gain frequency f_{max} but nevertheless supports high-order digital modulation such as the QAM [1,22]. Figure 4.83 shows its simplified block diagram and a

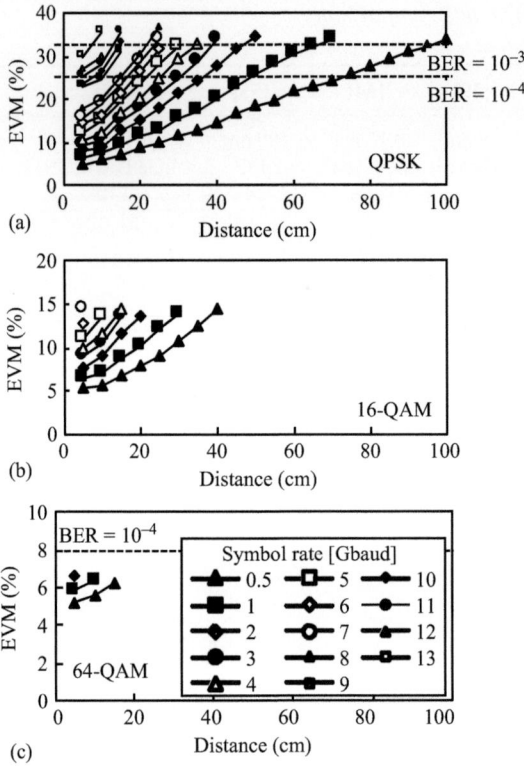

Figure 4.84   Measured EVM as a function of distance: (a) QPSK, (b) 16QAM, and
(c) 64QAM. Copyright 2016 IET. Reproduced, with permission,
from [25]

snapshot of a wireless digital data transmission experiment. The experimental data
presented in [1,22] were obtained mostly by directly connecting a measurement system
to the TX through a WG. In this section, we present the TX's wireless capability and
discuss how the relatively unexplored frequency band should be covered.

4.4.2   Wireless performance of 300-GHz CMOS transmitter

In the measurement setup, a WR3.4 WG probe leads to the transmitting horn antenna
as shown in Figure 4.83. The receiving horn antenna is mounted on a block downconverter (VDI WR3.4 MixAMC). The IF_1, centered at 18 GHz, was generated by
an arbitrary waveform generator (Keysight M8195A). The LO and RF frequencies
are 106 and 300 GHz, respectively. The frequencies of IF_1 and LO were chosen so
that unwanted image in IF_2 is reduced. We measured the EVM of the received signal
while changing the modulation format, the symbol rate, and the antenna-to-antenna
distance. A real-time oscilloscope (Keysight DSA-Z 334A) and an associated vector
signal analyzer software were used for the measurement. A channel equalizer built

Table 4.9 Performance summary of the 300-GHz CMOS transmitter

	QPSK		16QAM		64QAM	
Distance	1 m	5 cm	40 cm	5 cm	15 cm	5 cm
Data rate (Gbit/s)	1	26	2	28	3	12
EVM (%)	33.5	30.0	14.5	14.4	6.3	6.6

into the software was applied. The modulation formats used in the experiment were QPSK, 16QAM, and 64QAM. The measurement results are shown in Figure 4.84. The maximum symbol rate of 13 Gbaud was dictated by the bandwidth of an IF amplifier in the measurement system. Table 4.9 summarizes the longest distance and fastest transmission data for each modulation.

4.4.3 Comparison of transmit-receive systems

To put the experimental wireless performance of the 300-GHz CMOS TX into perspective, we introduce a FoM that allows comparison of transmit-receive systems with diverse configurations. Different systems have different TXs, RXs, antennas, distances, modulation formats, and symbol rates. We propose that all such differences but the symbol rate, r_s, be absorbed into an effective distance. For example, a wireless communication experiment performed using high-gain antennas can be interpreted to correspond to an experiment performed over a shorter effective distance with 0-dB i antennas. Similarly, a 64QAM experiment, which requires a high channel SNR, can be regarded as equivalent to a QPSK experiment, for which a lower SNR suffices, performed over a longer distance. Assuming that the antenna-to-antenna distance d is long enough (far-field assumption), we use the Friis transmission formula:

$$\frac{P_r}{P_t} = G_t G_r \left(\frac{\lambda}{4\pi d} \right)^2 \tag{4.10}$$

where P_t is the power fed into the transmitting antenna, G_t its antenna gain, P_r the power available from the receiving antenna, G_r its antenna gain, and λ the wavelength. From (4.10) follows

$$\left(\frac{d}{\sqrt{G_t G_r}} \right)^2 = \frac{P_t}{P_r} \left(\frac{\lambda}{4\pi} \right)^2 \tag{4.11}$$

The left-hand side can be regarded as an effective distance squared. To see how a difference in modulation should be incorporated, suppose that the modulation format is changed from a low-order one to a higher order one, which raises the requirement for channel SNR by a factor $R(>1)$. Then, the measured achievable distance d becomes shorter $(d/\sqrt{R})$. To make the effective distance invariant under the said simultaneous changes in modulation and distance $(d \to d/\sqrt{R})$, both sides of (4.11) should be multiplied by R. To be able to compare completely different systems, R should be defined using a fixed reference value, SNR_0, as $R \triangleq SNR/SNR_0$. We set $SNR_0 = 1$ and get

$$d_{\text{eff}}^2 \triangleq \left(d \sqrt{\frac{SNR}{G_t G_r}} \right)^2 = \frac{P_t \cdot SNR}{P_r} \left(\frac{\lambda}{4\pi} \right)^2 \tag{4.12}$$

where d_{eff} is our effective distance. It is the distance at which the channel SNR would become unity if a pair of 0-dB i antennas were used. Evaluation of SNR involves relating the measured BER to SNR. To do so, we make a simplifying assumption that the channel bandwidth B equals the r_s ($B = r_s$). Then, the information-bit-energy-to-noise-density ratio [60], $E_b/N_0 = SNR \cdot \log_2 M \cdot B/r_s = SNR \cdot \log_2 M$, where M is the signal order (number of points in a signal constellation). Put together with known theoretical relationships between BER and E_b/N_0 [60],

$$SNR = 2 \left[\text{erfc}^{-1} (2 \cdot BER) \right]^2 \quad (\text{OOK}) \tag{4.13}$$

$$SNR = \frac{2(M-1)}{3} \left[\text{erfc}^{-1} \left(\frac{BER \cdot \log_2 M}{2(1 - M^{-1/2})} \right) \right]^2 \quad (M\text{-QAM}) \tag{4.14}$$

where $\text{erfc}(x)$ the is complementary error function.

The signal bandwidth is proportional to the symbol rate r_s. The power spectral density, therefore, is inversely proportional to r_s. If the symbol rate is increased from a certain reference value r_{s0} to r_s, the SNR degrades and becomes (r_{s0}/r_s) times the original value. Dividing (4.12) by $(r_{s0}/r_s) \cdot (\lambda^2/r_{s0})$, we arrive at

$$FoM \triangleq \left(\frac{d_{\text{eff}}}{\lambda} \right)^2 r_s = \left(\frac{d}{\lambda} \right)^2 \frac{SNR}{G_t G_r} r_s \quad (\text{baud}) \tag{4.15}$$

The FoM represents the achievable symbol rate when $d_{\text{eff}} = \lambda$. The Friis formula (or the far-field assumption) breaks down at $d_{\text{eff}} = \lambda$ if d_{eff} is comparable to the actual distance d, but $d_{\text{eff}} \gg d$ is often satisfied.

Figure 4.85 plots FoMs of various experimental data on a symbol rate-vs-effective distance plane. Our TX, when combined with the measurement system, shows comparable FoM values to other 0.3-THz-or-higher systems [2,3,35,46,58,59]. In Figure 4.85, (4.15) gives a straight line of slope -2 for a given value of r_s. Our QPSK (o) and 16QAM ($\triangle$) results with $(d_{\text{eff}}/\lambda) \gtrsim 4$ appear to lie on such a line. Other, shorter distance results of ours are not on the line because d was too small and the far-field assumption was violated. A 240-GHz system employing high-performance GaAs mHEMTs [58] shows very high FoMs. To increase the FoM, the output power of the TX and/or r_s should be increased. To increase the output power at 300 GHz ($>f_{\max}$), where power amplification is not possible, our TX performs 32-way power combining [1].

Given the absence of any wireless communication standards at these frequencies, the upper bound value of r_s (vertical axis of Figure 4.85) is dictated by the hardware.

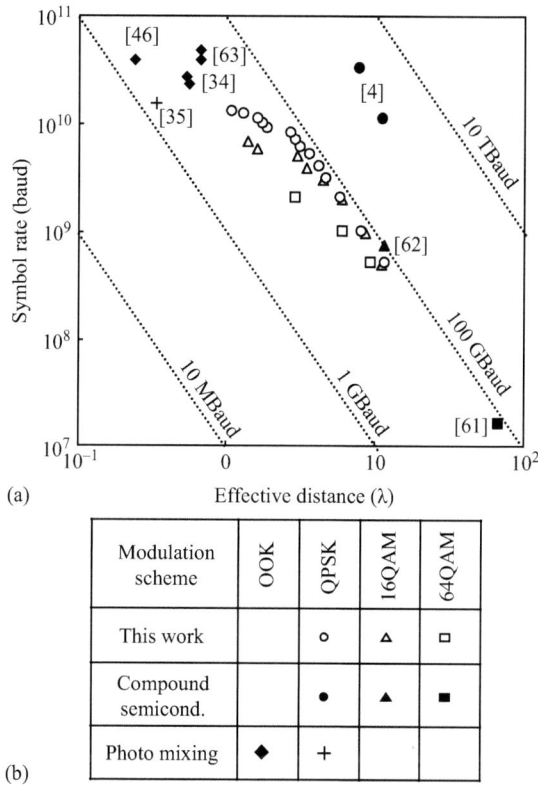

(a)

Modulation scheme	OOK	QPSK	16QAM	64QAM
This work		○	△	□
Compound semicond.		●	▲	■
Photo mixing	◆	+		

(b)

Figure 4.85   *Comparison of digital data transmission experiments: (a) FoMs on*
r_s-vs-d_{eff}/λ plane and (b) list of symbols. OOK, on-off keying.
Copyright 2016 IET. Reproduced, with permission, from [25]

In our case, r_s is limited by the bandwidth of the RX IF amplifier. If RX, too, is to be implemented with a CMOS technology with merely adequate terahertz performance, the maximum r_s is likely to become lower. In such a system, highest data rates ($r_s \log_2 M$) will be achieved by combining a moderate r_s with QAM (Table 4.9). This backs up the proposal in [1] that a 30-GHz-wide band be covered with six 5-GHz-wide QAM channels. If, on the other hand, high-performance devices are available, it seems appropriate to aim for extremely high data rates ($>50\,\text{Gb/s}$) with very high r_s and a simpler modulation, as in [56].

4.4.4   Conclusion

We presented the wireless performance of a 300-GHz CMOS TX [1,22]. The peak data rate reached 28 Gb/s with 16QAM (Figure 4.84 and Table 4.9). We introduced an FoM, (4.15), that allows comparison of transmit-receive systems

and considered how the vast frequency band around 300 GHz could be best utilized. If a CMOS technology is to be adopted, it is best to cover a very wide
bandwidth (several tens of GHz) with multiple QAM channels with a reasonable
per-channel data rate [1]. This underpins the importance of QAM-capability even at
300 GHz, where very wide bandwidths are available [1]. Superior integration capability of CMOS technology could facilitate possible eventual introduction of channel
bonding.

References

[1]   K. Katayama, K. Takano, S. Amakawa, *et al.*, "A 300 GHz 40 nm CMOS trans-
      mitter with 32-QAM 17.5 Gb/s/ch capability over 6 channels," *International
      Solid-State Circuits Conference*, pp. 342–343, Feb. 2016.

[2]   C. Jastrow, S. Priebe, B. Spitschan, *et al.*, "Wireless digital data transmission
      at 300 GHz," *Electronics Letters*, vol. 46, no. 9, pp. 661–663, 2010.

[3]   C. Wang, B. Lu, C. Lin, *et al.*, "0.34-THz wireless link based on high-
      order modulation for future wireless local area network applications," *IEEE
      Transactions on Terahertz Science and Technology*, vol. 4, no. 1, pp. 75–85,
      2014.

[4]   D. Lopez-Diaz, I. Kallfass, A. Tessmann, *et al.*, "A subharmonic chipset
      for gigabit communication around 240 GHz," *IEEE MTT-S International
      Microwave Symposium*, pp. 1–3, Jun. 2012.

[5]   Z. Wang, P.-Y. Chiang, P. Nazari, C.-C. Wang, Z. Chen, and P. Heydari,
      "A CMOS 210-GHz fundamental transceiver with OOK modulation," *IEEE
      Journal of Solid-State Circuits*, vol. 49, no. 3, pp. 564–580, 2014.

[6]   S. Kim, J. Yun, D. Yoon, *et al.*, "300 GHz integrated heterodyne receiver
      and transmitter with on-chip fundamental local oscillator and mixers," *IEEE
      Transactions on Terahertz Science and Technology*, vol. 5, no. 1, pp. 92–101,
      2015.

[7]   N. Sarmah, J. Grzyb, K. Statnikov, *et al.*, "A fully integrated 240-GHz direct-
      conversion quadrature transmitter and receiver chipset in SiGe technology,"
      IEEE Transactions on Microwave Theory and Techniques, vol. 64, no. 2,
      pp. 562–574, 2016.

[8]   M. Seo, M. Urteaga, A. Young, *et al.*, "A single-chip 630 GHz transmitter with
      210 GHz sub-harmonic PLL local oscillator in 130 nm InP HBT," *IEEE MTT-S
      International Microwave Symposium*, pp. 1–3, Jun. 2012.

[9]   H.-J. Song, J.-Y. Kim, K. Ajito, N. Kukutsu, M. Yaita, "50-Gb/s direct conver-
      sion QPSK modulator and demodulator MMICs for terahertz communications
      at 300 GHz," *IEEE Transactions on Microwave Theory and Techniques*, vol. 62,
      no. 3, pp. 600–609, 2014.

[10] S. Hu, Y.-Z. Xiong, B. Zhang, *et al.*, "A SiGe BiCMOS transmitter/
      receiver chipset with on-chip SIW antennas for terahertz applications," *IEEE
      Journal of Solid-State Circuits*, vol. 47, no. 11, pp. 2654–2664, 2012.

[11] J.-D. Park, S. Kang, S. V. Thyagarajan, E. Alon, A. M. Niknejad, "A 260 GHz fully integrated CMOS transceiver for wireless chip-to-chip communication," *Symposium on VLSI Circuits*, pp. 48–49, Jun. 2012.

[12] S. Kang, S. V. Thyagarajan, A. M. Niknejad, "A 240 GHz wideband QPSK transmitter in 65 nm CMOS," *Radio Frequency Integrated Circuits Symposium*, pp. 353–356, Jun. 2014.

[13] R. Dong, S. Hara, I. Watanabe, *et al.*, "Power spectrum analysis of a tripler-based 300-GHz CMOS up-conversion mixer," *European Microwave Conference*, pp. 345–348, Oct. 2016.

[14] S. A. Maas, *Nonlinear Microwave and RF Circuits*, 2nd edition, Artech House, Norwood, MA, 2003.

[15] R. Mavaddat, *Network Scattering Parameters*, World Scientific, Singapore, 1996.

[16] S. Amakawa, "Theory of gain and stability of small-signal amplifiers with lossless reciprocal feedback," *Asia-Pacific Microwave Conference*, pp. 1184–1186, Nov. 2014.

[17] S. Amakawa, Y. Ito, "Graphical approach to analysis and design of gain-boosted near-f_{max} feedback amplifiers," *European Microwave Conference*, pp. 1039–1042, Oct. 2016.

[18] P. A. Rizzi, *Microwave Engineering: Passive Circuits*, Prentice Hall, Upper Saddle River, NJ, 1988.

[19] R. W. P. King, *Transmission-Line Theory*, Dover, New York, 1965.

[20] L. N. Dworsky, *Modern Transmission Line Theory and Applications*, Wiley-Interscience, New York, 1979.

[21] G. Miano, A. Maffucci, *Transmission Lines and Lumped Circuits*, Academic Press, San Diego, CA, 2001.

[22] K. Katayama, K. Takano, S. Amakawa, S. Hara, T. Yoshida, M. Fujishima, "CMOS 300-GHz 64-QAM transmitter," *IEEE MTT-S International Microwave Symposium*, pp. 1–4, May 2016.

[23] K. Takano, K. Katayama, S. Amakawa, T. Yoshida, M. Fujishima, "A 300-GHz 64-QAM CMOS transmitter with 21-Gb/s maximum per-channel data rate," *European Microwave Integrated Circuits Conference*, pp. 193–196, Oct. 2016.

[24] K. M. Gharaibeh, K. G. Gard, M. B. Steer, "Accurate estimation of digital communication system metrics—SNR, EVM and ρ in a nonlinear amplifier environment," *64th Automatic RF Techniques Group (ARFTG) Conference*, pp. 41–44, Dec. 2004.

[25] K. Takano, K. Katayama, S. Amakawa, T. Yoshida, M. Fujishima, "Wireless digital data transmission from a 300 GHz CMOS transmitter," *Electronics Letters*, vol. 52, no. 15, pp. 1353–1355, 2016.

[26] F. Boes, T. Messinger, J. Antes, *et al.*, "Ultra-broadband MMIC-based wireless link at 240 GHz enabled by 64GS/s DAC," *International Conference on Infrared Millimeter and Terahertz Waves*, pp. 1–2, Sep. 2014.

[27] I. Kallfass, F. Boes, T. Messinger, *et al.*, "64 Gbit/s transmission over 850 m fixed wireless link at 240 GHz carrier frequency," *Journal of Infrared, Millimeter, and Terahertz Waves*, vol. 36, no. 2, pp. 221–233, 2015.

[28]   S. Zeinolabedinzadeh, M. Kaynak, W. Khan, *et al.*, "A 314 GHz, fully-integrated SiGe transmitter and receiver with integrated antenna," *Radio Frequency Integrated Circuits Symposium*, pp. 361–364, Jun. 2014.

[29]   I. Kallfass, I. Dan, S. Rey, *et al.*, "Towards MMIC-based 300 GHz indoor wireless communication systems," *IEICE Transactions on Electronics*, vol. E98-C, no. 12, pp. 1081–1090, 2015.

[30]   M. Fujishima, "Channel allocation of 300 GHz band for fiber-optic-speed wireless communication," *URSI Asia-Pacific Radio Science Conference*, pp. 330–333, Aug. 2016.

[31]   K. Takano, K. Katayama, S. Amakawa, T. Yoshida, M. Fujishima, "56-Gbit/s 16-QAM wireless link with 300-GHz-band CMOS transmitter," *IEEE MTT-S International Microwave Symposium*, Honolulu, HI, pp. 1–4, 2017.

[32]   M. Fujishima, "Terahertz wireless communication using 300 GHz CMOS transmitter," *IEEE 13th International Conference on Solid-State and Integrated Circuit Technology*, Oct. 2016.

[33]   T. Tajima, H.-J. Song, M. Yaita, "Design and analysis of LTCC-integrated planar microstrip-to-waveguide transition at 300 GHz," *IEEE Transactions on Microwave Theory and Techniques*, vol. 64, no. 1, pp. 106–114, 2016.

[34]   G. Ducournau, P. Szriftgiser, F. Pavanello, *et al.*, "THz communications using photonics and electronic devices: the race to data-rate," *Journal of Infrared, Millimeter, and Terahertz Waves*, vol. 36, no. 2, pp. 198–220, 2015.

[35]   G. Ducournau, P. Szriftgiser, A. Beck, *et al.*, "Ultrawide-bandwidth single-channel 0.4-THz wireless link combining broadband quasi-optic photomixer and coherent detection," *IEEE Transactions on Terahertz Science and Technology*, vol. 4, no. 3, pp. 328–337, 2014.

[36]   M. Fujishima, S. Amakawa, "Integrated-circuit approaches to THz communications: challenges, advances, and future prospects," *IEICE Transactions on Fundamentals of Electronics Communications and Computer Sciences*, vol. E100-A, no. 2, pp. 516–523, 2017.

[37]   M. K. Matters-Kammerer, L. Tripodi, R. van Langevelde, J. Cumana, R. H. Jansen, "RF characterization of Schottky diodes in 65-nm CMOS," *IEEE Transactions on Electron Devices*, vol. 57, no. 5, pp. 1063–1068, 2010.

[38]   D. Shim, K. K. O, "Self-biased anti-parallel diode pair in 130-nm CMOS," *Electronics Letters*, vol. 52, no. 13, pp. 1147–1149, 2016.

[39]   E. Öjefors, B. Heinemann, U. R. Pfeiffer, "Subharmonic 220- and 320-GHz SiGe HBT receiver front-ends," *IEEE Transactions on Microwave Theory and Techniques*, vol. 60, no. 5, pp. 1397–1404, 2012.

[40]   K. Takano, S. Amakawa, K. Katayama, *et al.*, "A 105Gb/s 300GHz CMOS Transmitter," *International Solid-State Circuits Conference*, pp. 308–309, Feb. 2017.

[41]   Keysight Technologies, "High-accuracy noise figure measurements using the PNA-X series network analyzer," *Application Note 1408-20*, Retrieved November 7, 2016, from http://cp.literature.agilent.com/litweb/pdf/5990-5800EN.pdf

[42] S. Hara, K. Katayama, K. Takano, *et al.*, "A 416-mW 32-Gbit/s 300-GHz CMOS receiver," *IEEE International Symposium on Radio-Frequency Integration Technology*, pp. 65–67, 2017.

[43] M. Tytgat, M. Steyaert, P. Reynaert, "A 200 GHz downconverter in 90 nm CMOS," *European Solid-State Circuits Conference*, pp. 239–242, Sep. 2011.

[44] S. V. Thyagarajan, S. Kang, A. M. Niknejad, "A 240 GHz fully integrated wide-band QPSK receiver in 65 nm CMOS," *IEEE Journal of Solid-State Circuits*, vol. 50, no. 10, pp. 2268–2280, 2015.

[45] S. Hara, K. Katayama, K. Takano, *et al.*, "A 32 Gbit/s 16QAM CMOS receiver in 300 GHz band," *IEEE International Microwave Symposium*, pp. 1–4, Jun. 2017.

[46] H.-J. Song, K. Ajito, Y. Muramoto, A. Wakatsuki, T. Nagatsuma, N. Kukutsu, "24 Gbit/s data transmission in 300 GHz band for future terahertz communications," *Electronics Letters*, vol. 48, no. 15, pp. 953–954, 2012.

[47] T. Nagatsuma, S. Hisatake, H. H. Nguyen Pham, "Photonics for millimeter-wave and terahertz sensing and measurement," *IEICE Transactions on Electronics*, vol. E99-C, no. 2, pp. 173–180, 2016.

[48] S. Hara, K. Takano, K. Katayama, *et al.*, "300-GHz CMOS Receiver Module with WR-3.4 Waveguide Interface," *48th European Microwave Conference*, pp. 396–399, 2018.

[49] T. Kürner, "What's next? Wireless communication beyond 60 GHz", *Tutorial of IEEE 802.15 THz Interest Group at IEEE 802 Plenary*, Jul. 2012.

[50] H.-J. Song, "Packages for terahertz electronics," *Proceedings of the IEEE*, vol. 105, pp. 1121–1138, 2017.

[51] K. Takano, K. Katayama, S. Hara, *et al.*, "300-GHz CMOS transmitter module with built-in waveguide transition on a multilayered glass epoxy PCB," *IEEE Radio Wireless Symposium*, pp. 154–156, Jan. 2018.

[52] N. M. Ridler, R. A. Ginley, "A review of the IEEE 1785 standards for rectangular waveguides above 110 GHz," *89th ARFTG Microwave Measurement Conference*, pp. 1–4, Jun. 2017.

[53] S. Lee, R. Dong, T. Yoshida, *et al.*, "An 80Gb/s 300GHz-Band Single-Chip CMOS Transceiver," *International Solid-State Circuits Conference*, pp. 170–171, 2019.

[54] P. Rodríguez-Vázquez, N. Sarmah, B. Heinemann, U. Pfeiffer, "A 65 Gbps QPSK one meter wireless link operating at a 225–255 GHz tunable carrier in a SiGe HBT technology," *IEEE Radio Wireless Symposium*, pp. 146–149, Jan. 2018.

[55] H. Hamada, T. Fujimura, I. Abdo, *et al.*, "300-GHz 100-Gb/s InP-HEMT wireless transceiver using a 300-GHz fundamental mixer," *IEEE International Microwave Symposium*, pp. 1480–1483, Jun. 2018.

[56] J. Al-Eryani, H. Knapp, J. Kammerer, K. Aufinger, H. Li, L. Maurer, "Fully integrated single-chip 305–375-GHz transceiver with on-chip antennas in SiGe BiCMOS," *IEEE Transactions on Terahertz Science and Technology*, vol. 8, no. 3, pp. 329–339, 2018.

[57]   J. S. Engberg, G. B. Gawler, "Significance of the noise measure for cascaded stages," *IEEE Transactions on Circuit Theory*, vol. 16, no. 2, pp. 259–260, 1969.

[58]   G. Ducournau, K. Engenhardt, P. Szriftgiser, *et al.*, "32 Gbit/s QPSK transmission at 385 GHz using coherent fibre-optic technologies and THz double heterodyne detection," *Electronics Letters*, vol. 51, no. 12, pp. 915–917, 2015.

[59]   T. Nagatsuma, S. Horiguchi, Y. Minamikata, *et al.*, "Terahertz wireless communications based on photonics technologies," *Optics Express*, vol. 21, no. 20, pp. 23736–23747, 2013.

[60]   E. McCune, *Practical Digital Wireless Signals*, Cambridge University Press, Cambridge, 2010

[61]   R. Piesiewicz, T. Kleine-Ostmann, N. Krumbholz, *et al.*, "Short-range ultra-broadband terahertz communications: concepts and perspectives," *IEEE Antennas and Propagation Magazine*, vol. 49, no. 6, pp. 24–39, 2007.

[62]   K.-C. Huang, Z. Wang, "Terahertz terabit wireless communication," *IEEE Microwave Magazine*, vol. 12, no. 4, pp. 108–116, 2011.

[63]   A. Hirata, M. Yaita, "Ultrafast terahertz wireless communications technologies," *IEEE Transactions on Terahertz Science and Technology*, vol. 5, no. 6, pp. 1128–1132, 2015.

Chapter 5

Future prospects

We have discussed circuit technologies and examples of architectures that are needed to realize complementary metal oxide semiconductor (CMOS) transceivers in the terahertz band. At the time of writing this book, there are still several issues to be solved until the terahertz CMOS circuit matures to technology that anyone can use. In the circuit, issues to be addressed such as a low-noise signal source and a high-speed baseband circuit remain. It is necessary to take time to study these issues in the future. However, on the other hand, the discussion on what kind of future the terahertz communication system opens will be helpful in considering what kind of technology should be developed. In the last chapter, we would like to discuss how to use frequencies in the 300 GHz band, and what kind of applications will be expanded in future if technology matures.

5.1 Channel allocation planning for 300-GHz band [8]

What are the main performance requirements and practical considerations for THz radio [1]? First of all, the highest data rate ought to be significantly higher than that of the existing 60-GHz wireless standards. In addition, frequency band must be chosen such that the atmospheric losses are low if it is to be used both indoors and outdoors. Furthermore, some compatibility with the 60-GHz band should be kept so that the baseband circuitry developed for that band can be reused at least to a degree.

Let us first consider the choice of a frequency band. THz radio should use the unallocated frequencies above 275 GHz. What about the upper bound frequency? There is an absorption peak due to water vapor at 325 GHz. The atmospheric loss at 325 GHz is about 28 dB/km as shown in Figure 5.1. If THz radio is to be used over kilometer-level distances, frequencies around 325 GHz should be avoided. If we choose to use the lower side of this absorption peak, the upper bound frequency would be around 320 GHz. To consider what the lower bound frequency should be, note that the frequency range from 252 to 275 GHz has already been allocated for fixed or mobile radio communications (Fig. 1.11) [2]. This range of frequencies could be used together with those above 275 GHz. Then, the lower bound frequency would be 252 GHz.

Standard rectangular waveguide specifications should also be taken into account. At frequencies above 110 GHz, rectangular waveguides are most commonly used.

A waveguide of given dimensions have nominal usable lower and upper bound frequencies. Crossing those frequency boundaries would require two different sets of measurement equipment and is highly undesirable. The WR3.4 waveguide covers the frequency range from 220 to 325 GHz, and the aforementioned frequency range of 252–320 GHz is completely covered by it.

Another consideration is the compatibility with existing wireless standards. IEEE 802.11ad defines the 60-GHz band wireless local area network (WLAN) covering 57–66 GHz. There are four 2.16-GHz-wide channels in this frequency range as shown in Figure 5.2. The center frequencies of these channels are all integer multiples of 2.16 GHz. To allow the use of local oscillator circuits developed for the 60-GHz band and also for the 300-GHz band, the channel frequencies in the latter should also be integer multiples of 2.16 GHz.

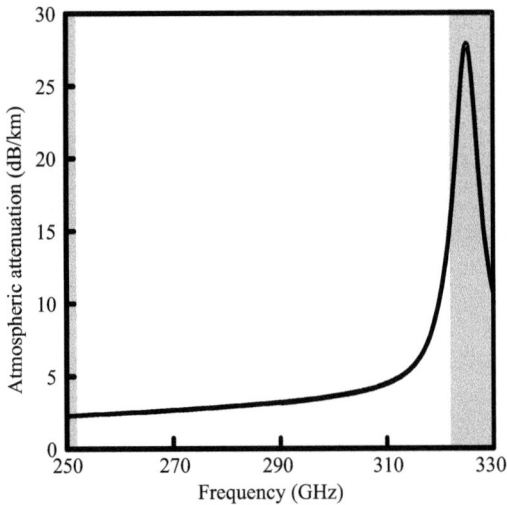

Figure 5.1   *Atmospheric attenuation above 250 GHz [3]. Absorption by water vapor appears at 325 GHz. Copyright 2017 IEICE. Reproduced, with permission, from [9]*

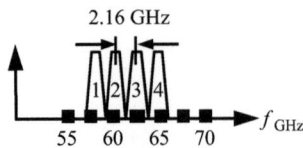

Figure 5.2   *The channel allocation for the 60-GHz-band WLAN according to IEEE 802.11ad. The center frequencies are 58.32, 60.48, 62.64, and 64.80 GHz. Copyright 2017 IEICE. Reproduced, with permission, from [9]*

While compatibility is important, extensibility is yet another consideration. IEEE 802.11ay, currently under development, aims to extend the data rate through bonding two or four 2.16-GHz-wide channels of 802.11ad. The whole purpose of using the 300-GHz band is to make good use of the vast available bandwidth and achieve very high data rates not achievable using the 60-GHz band.

In IEEE 802.15.3d, the standard for channel allocation in the 300 GHz band has been established [4]. In this standard, channels are allocated to frequency bands close to 70 GHz from 252 to 321 GHz (Figure 5.3), which are composed of frequencies already allocated for communication and frequencies expected to be allocated to communication in the future. To maintain compatibility with the conventional 60 GHz band standard, a plurality of channel bands that are multiples of 2.16 GHz, as shown in Figure 5.3(a) and (b), are allocated to this frequency band. On the other hand, considering that the data rate of wireless communication is close to optical communication, it is preferable to consider compatibility with wired communication standards. For example, the 28 Gbit/s standard is widely used in NRZ (nonreturn to zero) transmission called CEI-28G (common electrical interface 28 Gbit/s) for electrical interfaces

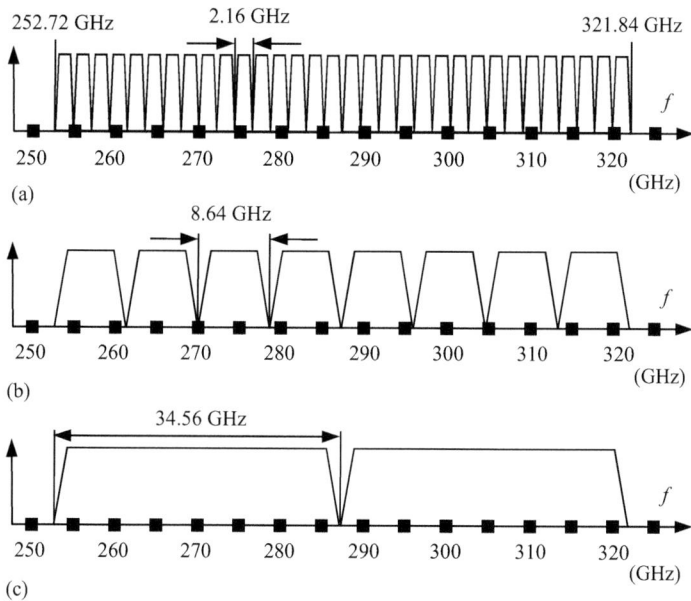

Figure 5.3 *In IEEE 802.15.3d, channels in multiple bands are allocated in the frequency band from 252.72 to 321.84 GHz. (a) and (b) are examples of channels allocated in IEEE 802.15.3d. (c) Assigning the channel of the 34.56 GHz band facilitates connection with the wired communication interface CEI-28G (Common Electrical Interface 28 Gigabit-per-second). Copyright 2017 IEICE. Reproduced, with permission, from [9]*

of field programmable gate arrays and optical fiber. Assigning these two NRZ signals to the I/Q signals used for quadrature phase shift keying (QPSK) results in 56 Gbit/s wireless communication. Since the symbol rate is 28 Gbaud (gigabaud), when it is modulated with a roll-off coefficient of 0.2, it becomes a bandwidth of 33.4 GHz. According to the IEEE 802.15.3d standard, a channel suitable for this band is not currently defined. However, allocating 34.56 GHz, which is 16 times 2.16 GHz, to one channel is suitable for connecting the 28 Gbit/s NRZ wired signal to the 56 Gbit/s QPSK radio signal. As long as the channel bandwidth is 34.56 GHz, as shown in Figure 5.3(c), two channels can be allocated to the entire frequency band specified by IEEE 802.15.3d. Assigning channels in this way, the channel on the low frequency side from 252 to 287 GHz has a long distance of about 2 km as the atmospheric attenuation becomes 10 dB, which is suitable for medium-to-long-distance communication.

5.2   The future of terahertz communication spreading in space [5]

Inter-server communication for data centers, backhaul and fronthaul of networks, and high-speed communication in electronic devices are assumed applications of terahertz communication, but these are forms of fixed wireless communication that have been realized by the conventional optical fiber network. Merely replacing fixed communication, however, does not achieve the maximum potential of terahertz communication. On the other hand, since terahertz air attenuation is large, it may be thought that it is unsuitable for long-distance communication. However, as mentioned in Section 1.2, there is an atmosphere window in the 300 GHz band. In addition, if the antenna size is maintained, even if the wavelength is short, the directivity increases, but the propagation loss decreases. The received power increases as the frequency increases if the effective area of the antenna is constant, so ultrahigh-speed terahertz communication may be utilized in the universe without atmospheric attenuation if a high-power terahertz transmitter and a high-gain antenna are available. In the 300 GHz band, the gain of a parabola antenna with a 1.2-m diameter will be 70 dBi from (1.3) if the aperture efficiency is 70%. The diameter ϕ of the beam produced by the antenna with the gain G at the distance d is given by

$$\phi \simeq \frac{4d}{\sqrt{G}} \tag{5.1}$$

regardless of the frequency. With a 70 dBi antenna, the diameter of the beam 1,000 km ahead will be 1.3 km. Terahertz communication in which the beam diameter is wider than the laser beam helps alleviate the requirement of position control of the antenna. Furthermore, in the terahertz communication in the universe without atmospheric attenuation, a frequency band exceeding 300 GHz can be utilized. Although not in the 300 GHz band, as shown in Figure 1.21, in the frequency band from 121 to 154 GHz, the distance at which the atmospheric attenuation becomes 10 dB exceeds 8 km. If this wide frequency bandwidth that is not currently allocated to communication is available, stable high-speed communication to the universe is possible in the desert without rain attenuation located at an altitude of 5,000 m at which the air density is about half that of the sea level.

In the future, can we use terahertz communication to solve social problems? For example, to simulate various phenomena in the real world accurately on a large scale, it is indispensable to improve the performance of supercomputers. What are the challenges that must be solved when performance improvement of supercomputers is expected in the future? The computing performance of the supercomputer has improved exponentially, by 1,000 times in about 10 years [6]. Currently, development is progressing toward the realization of an exascale computer that performs floating-point calculations 10^{18} times per second in 2020. If the computation speed continues to improve after 2020, a zettascale computer will be created in 2030 that will perform floating point calculations at a tremendous rate of 10^{21} times per second. One problem here is power consumption. It is foreseen that the power consumption of an exascale computer will be 20 MW [7]. On the other hand, in a zettascale computer with 1,000 times higher computing capacity, the power consumption will be 20 GW if the hardware scale is simply multiplied by 1,000. Even if low-power technology is developed, the power consumption of the zettascale computer will be around 1 GW. The power consumption of 1 GW is roughly equivalent to the output of one nuclear power generator. How can we maintain a sustainable society while covering the enormous power consumption of such supercomputers? Research on space photovoltaic power generation using a 2.5-km^2 solar panel on a geostationary orbit is proceeding as one of stable candidates for renewable energy [8]. This solar panel can generate 1 GW. Electricity produced by space photovoltaic power will be transmitted to the ground using microwaves. The target efficiency of power transmission is 50%. If a supercomputer is built together with the space photovoltaic panel, and electric power generated by space photovoltaic power is used in place, power can be effectively utilized without transmission loss [8]. Ultrahigh-speed terahertz communication can be used for large-volume data exchange for space supercomputers that cannot lay optical fiber.

On the other hand, it will be an era of monitoring the real world with enormous sensors exceeding 1 trillion units and processing that information with artificial intelligence. Although the information obtained from the sensor cannot be stored as it is, the amount of stored data will exponentially increase. On the other hand, cold data in which access frequency decreases with time will also increase. Since the access frequency of cold data is low, the access speed is not important. When slow access is allowed, it may be possible to construct a cold storage data center on the moon, for example, rather than on Earth [9]. Since the light propagation time to the moon is 1.3 s, the access speed will be about 3 s in a round trip. Nonetheless, a lunar cold storage data center will be one of the candidates for utilization on the moon.

Furthermore, manned exploration of Mars and migration of mankind are currently being studied [10]. The means of communication of the migrants to Mars is limited to radio. It is noted that 1 PB can be transmitted per day when information can be sent at a data rate of 100 Gbit/s. When Mars is furthest from Earth it takes 21 min to propagate electromagnetic waves. Hence, although communication will not be real time, terahertz communication makes it possible to exchange large volumes of

Required output power and antenna diameter

	Output power	Antenna diameter
Geostationary orbit	1 W	4.5 m
Moon	10 W	8 m
Mars	100 W	130 m

100 Gb/s (16 QAM/25 Gbaud) at 300 GHz
Receiver noise figure: 10 dB
Parabola antenna with aperture efficiency of 0.6
Bit error rate: 10^{-3}

Mars

380,000,000 km (max.)

Geostationary orbit

36,000 km

Moon

380,000 km

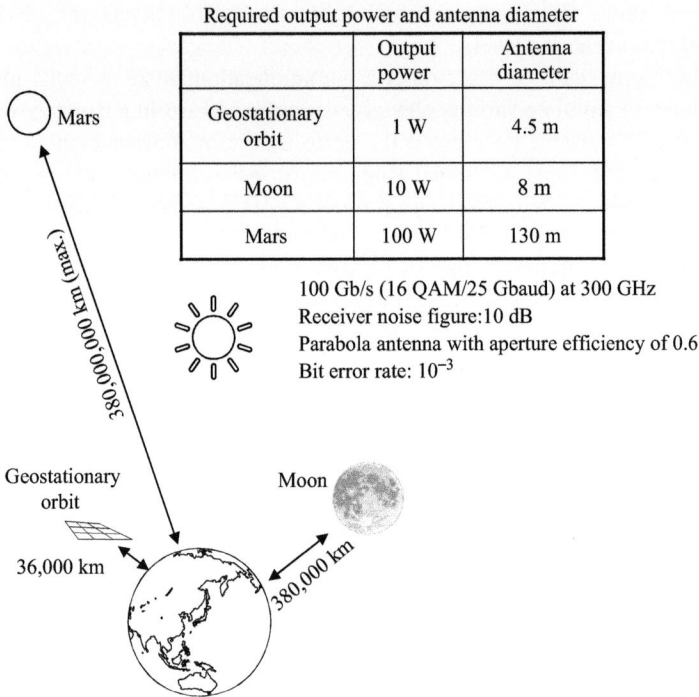

Figure 5.4   Estimates of the performance required when 300 GHz band communication is realized in space. If the transmission power of terahertz communication increases, a broadband communication link toward space becomes possible

information between Earth and Mars. It is noted that the moon's atmosphere is very low (i.e., almost vacuum). The atmospheric pressure of Mars is about 0.6% of the Earth, and most of the atmosphere is carbon dioxide. Carbon dioxide absorbs infrared to cause a greenhouse effect but does not absorb terahertz. Therefore, terahertz absorption on the moon and Mars is negligible. Performance required for achieving terahertz communication with geostationary orbits of the moon and Mars is estimated in Figure 5.4. Although it may not be a technology that can be realized immediately, the space utilization of ultrahigh-speed terahertz communication will be a meaningful challenge to overcome for the sustainable development of mankind.

5.3   Summary

The 300-GHz band has an acceptable atmospheric loss of 10 dB/km and is the last remaining frequency band with a vast bandwidth of nearly 70 GHz. 300-GHz-band wireless links with fiber-optic speeds could even affect space programs in the 2020s and beyond. Currently, the performance of IC-based THz circuits is not sufficient. Fiber-optic data rates also require ultrahigh-speed baseband signal processing, and for that, advanced CMOS technology is necessary. On the other hand, high-performance compound semiconductor technology is more suitable for transmitter power amplifiers (PAs) and receiver low-noise amplifiers (LNAs). For the radio-transceiver (RF) front-end core that sits in between PA/LNA and the baseband circuitry, CMOS technology might be suitable because of its superior integration capability. In any case, for successful deployment of THz wireless, the best mix of CMOS and compound-semiconductor technologies will be needed. For applications that require Watt-level output power, traveling-wave tubes [11] might be needed. Heterogeneous integration of advanced CMOS, high-performance compound semiconductors, and vacuum tubes could be a key to opening up the future of ultrahigh-speed wireless communications and applications that need them.

5.4 Concluding remarks

Two hundred years has passed and communication technology has developed dramatically since the first working telegraph with static electricity was built in 1816 by Francis Ronalds, a British inventor. In 1920, Barkhausen–Kurz tube was invented by German physicists Heinrich Georg Barkhausen and Karl Kurz. As a result, radio wave with ultrahigh frequency exceeding 300 MHz can be generated, and microwave research will become popular. Fifty years later, in 1970, Corning Glass Works developed a low attenuation optical fiber for communication purposes. From this point on, research on fiber-optic communication becomes popular. Large-capacity and long-distance transmission has become possible by fiber-optic communication. However, the microwave, which is wireless, is not replaced by fiber-optic communication that is wired. Fiber-optic communication provided applications that could not be realized with microwaves, and microwaves has been never lost. Meanwhile, 2020 will be coming soon, which will be just 50 years from 1970 when the optical fiber was developed. As shown in Figure 5.5, human beings acquire terahertz as a new communication medium, and research on terahertz communication will become more

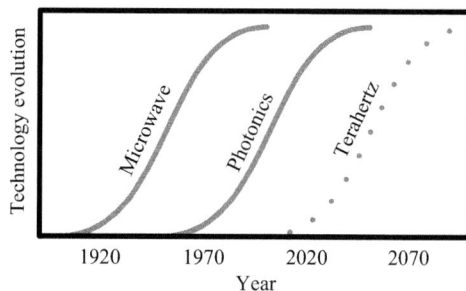

Figure 5.5   The leading role in research and development of communication media has changed every 50 years

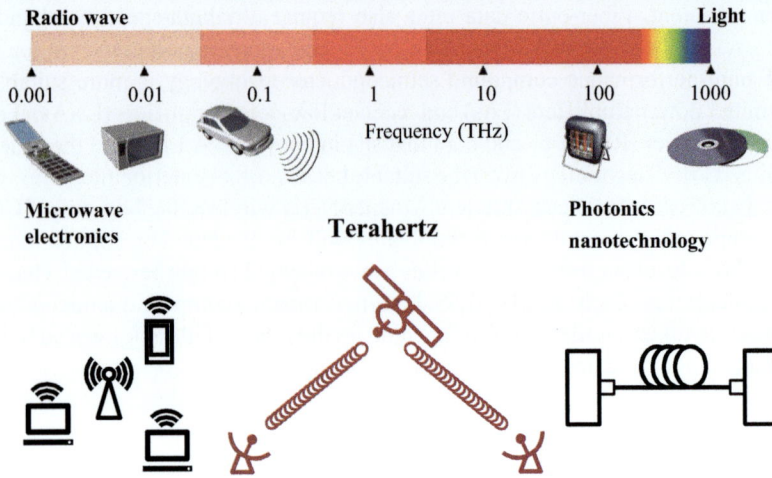

Figure 5.6   Terahertz not only fills the gap between the wavelengths of radio waves but also fills the application gap of microwave and optical fiber communication

popular in the future. Even if terahertz communication is put into practical use, neither fiber-optic communication nor microwave communication will be lost. Just as terahertz fills the wavelength between radio waves and light as shown in Figure 5.6, we believe that terahertz fills the application gap between microwave and optical fiber communication.

References

[1]   M. Fujishima, "Channel allocation of 300GHz band for fiber-optic-speed wireless communication," *URSI Asia-Pacific Radio Science Conference*, pp. 330–333, Aug. 2016.
[2]   International Telecommunication Union. Article 5: Frequency allocations. In Radio Regulations Article Edition of 2012; International Telecommunication Union (ITU): Geneva, Switzerland, 2012; pp. 37–178. Available online: http://search.itu.int/history/HistoryDigitalCollectionDocLibrary/1.41.48.en.101.pdf
[3]   International Telecommunication Union. Recommendation ITU-R P.676-11. In Attenuation by Atmospheric Gases; P Series Radiowave Propagation; International Telecommunication Union (ITU): Geneve, Switzerland, 2016; Available online: https://www.itu.int/dms_pubrec/itu-r/rec/p/R-REC-P.676-11-201609-I!!PDF-E.pdf

[4] *IEEE Standard for High Data Rate Wireless Multi-Media Networks, Amendment 2: 100 Gb/s Wireless Switched Point-to-Point Physical Layer*, IEEE Computer Society sponsored by the LAN/MAN Standards Committee. Available online: https://standards.ieee.org/standard/802_15_3d-2017.html.

[5] M. Fujishima, "Key Technologies for THz Wireless Link by Silicon CMOS Integrated Circuits," *Photonics*, vol. 5, no. 4, pp. 1–17, 2018.

[6] Performance development. Available online: https://www.top500.org/statistics/perfdevel/

[7] P. Kogge, K. Bergman, S. Borkar, *et al.*, "Exascale computing study: Technology challenges in achieving exascale systems," *DARPA IPTO Report*, 2008. Available online: http://www.cse.nd.edu/Reports/2008/TR-2008-13.pdf

[8] M. Mori, H. Kagawa, Y. Saito, "Summary of studies on space solar power systems of Japan Aerospace Exploration Agency (JAXA)," *Acta Astronautica*, vol. 59, no. 1–5, pp. 132–138, 2006.

[9] M. Fujishima, S. Amakawa, "Integrated-circuit approaches to THz communications: Challenges, advances, and future prospects," *IEICE Transactions on Fundamentals of Electronics, Communications and Computer Sciences*, vol. 100, no. 2, pp. 516–523, 2017.

[10] G. Daines, *NASA's Journey to Mars*, NASA, 2015. Available online: www.nasa.gov/content/nasas-journeyto-mars.

[11] N. Masuda, M. Yoshida, Y. Fujishita, J. Kobayashi, N. Sekine, A. Sugano, "Development of 0.1/0.3 THz-band traveling wave tubes," *IEICE Tech. Rep., ED2015-110*, 2015 (in Japanese).

Index